ELIZ [illegible] ce in Mathemati [illegible] x and subseque [illegible] ificate of Educa [illegible] ndon. She went [illegible] ndon comprehen [illegible] ge of the Mathe [illegible] ent. She now lectures in Mathematics at a College of Further Education: Kilburn Polytechnic.

key facts

Worked Examples

APPLIED MATHEMATICS, E. M. Peet, B.Sc.

BIOLOGY, A. G. Toole, B.Sc.

CHEMISTRY, J. E. Chandler, B.Sc.
and R. J. Wilkinson, Ph.D.

ECONOMICS, R. Maile, B.A.

GEOGRAPHY, C. Lines, M.Sc.

PHYSICS, G. M. George, B.Sc.

PURE MATHEMATICS, R. A. Parsons, B.Sc.
and A. G. Dawson, B.Sc.

PURE AND APPLIED MATHEMATICS,
R. A. Parsons, B.Sc. and A. G. Dawson, B.Sc.

GCE A-Level

Applied Mathematics

E. M. Peet, B.Sc.

key facts

Published by Charles Letts Books Ltd
London, Edinburgh and New York

Published 1983 by Charles Letts Books Ltd
Diary House, Borough Road, London SE1 1DW

1st edition, 1st impression

Made and printed by Charles Letts (Scotland) Ltd
ISBN 0 85097 551 4

Contents

Introduction

The contents of this book cover the majority of topics listed by the syllabuses of the various A-level Examination Boards. Each chapter consists of worked examples of a typical A-level standard on the topic under consideration.

An adequate preparation for the A-level examination will include the working of many past examination questions and students are advised to obtain and work through the past papers of the relevant Board.

Where questions have been reproduced verbatim from the past papers of A-level Examination Boards, they have been acknowledged by the appropriate initials of the Board:

AEB	The Associated Examining Board
SUJB	Southern Universities Joint Board
WJEC	Welsh Joint Education Committee

The Examination Boards accept no responsibility whatsoever for the accuracy or method of working in the answers given.

Notation

When, in the text, it is necessary to distinguish between vectors and scalars, vectors are denoted in bold roman type, and scalars in italic type, e.g. velocity **v**, speed *v* (or ***v***).

Chapter 1
Vectors

Example 1

1. A plane P_1 with normal vector $\mathbf{i}+\mathbf{j}-\mathbf{k}$ contains the point A with position vector $2\mathbf{i}-\mathbf{k}$ relative to an origin O. Find a vector equation for P_1.

A second plane P_2 has equation $\mathbf{r}.(2\mathbf{i}-\mathbf{j}+\mathbf{k})=0$. Find the angle between the two planes P_1 and P_2. Verify that the line of intersection, L, of the planes P_1 and P_2 passes through the point B, with position vector $\mathbf{i}+2\mathbf{j}$ relative to O, and is parallel to the vector $\mathbf{j}+\mathbf{k}$. Write down the vector equation of L and find the position vector of a point C on L such that AC is perpendicular to L. Find also the area of the triangle ABC.

(A.E.B.)

The standard form of the vector equation of a plane is $\mathbf{r}.\hat{\mathbf{n}}=d$ where $\mathbf{r}$ is the position vector of any point on the plane, $\hat{\mathbf{n}}$ is the unit vector perpendicular to the plane and d is the distance of the plane from the origin.

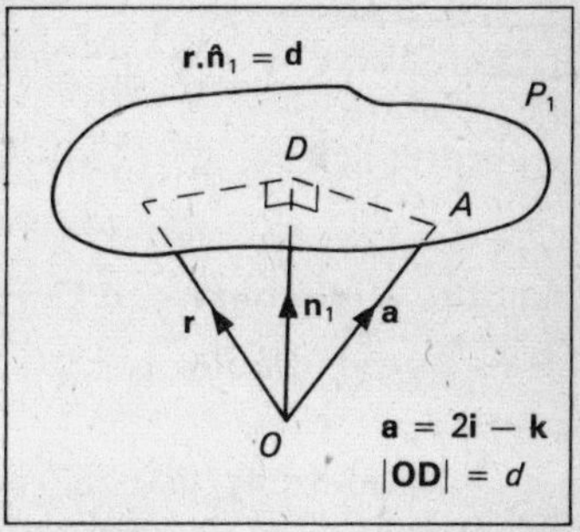

Figure 1

For plane P_1:

$$\hat{\mathbf{n}}_1=\frac{\mathbf{i}}{\sqrt{3}}+\frac{\mathbf{j}}{\sqrt{3}}-\frac{\mathbf{k}}{\sqrt{3}}$$

and $$d=(2\mathbf{i}-\mathbf{k}).\left(\frac{\mathbf{i}}{\sqrt{3}}+\frac{\mathbf{j}}{\sqrt{3}}-\frac{\mathbf{k}}{\sqrt{3}}\right)=\sqrt{3}$$

Hence, **a vector equation for P_1** is

$$\mathbf{r}.\left(\frac{\mathbf{i}}{\sqrt{3}}+\frac{\mathbf{j}}{\sqrt{3}}-\frac{\mathbf{k}}{\sqrt{3}}\right)=\sqrt{3}$$

or $\mathbf{r}.(\mathbf{i}+\mathbf{j}-\mathbf{k})=3$

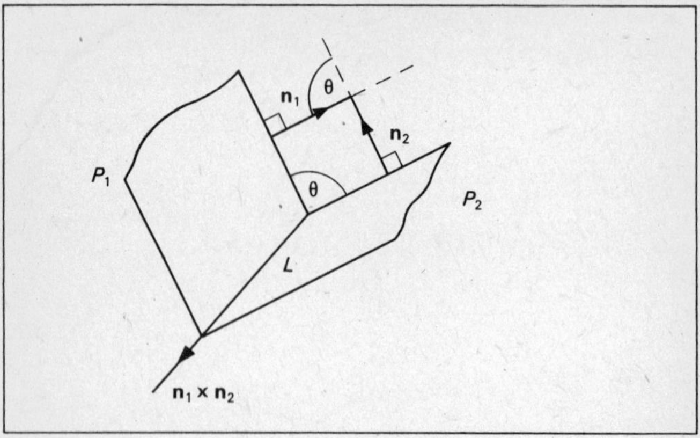

Figure 2

The angle between the two planes P_1 and P_2 is equal to the angle between the normals to the planes.

Refer to figure 2.

$$\cos\theta = \hat{\mathbf{n}}_1 . \hat{\mathbf{n}}_2 = \frac{1}{\sqrt{3}}(\mathbf{i}+\mathbf{j}-\mathbf{k}) . \frac{1}{\sqrt{6}}(2\mathbf{i}-\mathbf{j}+\mathbf{k}) = 0$$

and **the angle between the planes is 90°, i.e.**
P_1 is perpendicular to P_2.

In order to verify that the point B lies on the line of the intersection, L, of the planes P_1 and P_2, we need to show that $\mathbf{b} = \mathbf{i} + 2\mathbf{j}$ satisfies the vector equations of both P_1 and P_2.

$$P_1: \quad \mathbf{r}.(\mathbf{i}+\mathbf{j}-\mathbf{k}) = 3$$

Since $(\mathbf{i}+2\mathbf{j}).(\mathbf{i}+\mathbf{j}-\mathbf{k}) = 2+1=3$, then $\mathbf{b}$ lies in P_1.

$$P_2: \quad \mathbf{r}.(2\mathbf{i}-\mathbf{j}+\mathbf{k}) = 0$$

Since $(\mathbf{i}+2\mathbf{j}).(2\mathbf{i}-\mathbf{j}+\mathbf{k}) = 2-2=0$, then $\mathbf{b}$ lies in P_2.

Hence B lies on L.

Since the line of intersection is contained in both planes, then it is perpendicular to both $\mathbf{n}_1$ and $\mathbf{n}_2$, i.e. it is parallel to $\mathbf{n}_1 \times \mathbf{n}_2$.

$$\mathbf{n}_1 \times \mathbf{n}_2 = \begin{vmatrix} \mathbf{i} & \mathbf{j} & \mathbf{k} \\ 1 & 1 & -1 \\ 2 & -1 & 1 \end{vmatrix} = (1-1)\mathbf{i} - (1+2)\mathbf{j} + (-1-2)\mathbf{k}$$

$$= -3(\mathbf{j}+\mathbf{k})$$

Hence **the line of intersection is parallel to $(\mathbf{j}+\mathbf{k})$.**

A vector equation for L of the form '$\mathbf{r}=\mathbf{p}+\lambda\mathbf{q}$' may be found where $\mathbf{r}$ is the position vector of any point on the line, $\mathbf{p}$ is the position vector of a particular point on the line and $\mathbf{q}$ is a direction vector for the line. λ is a scalar constant.

$$L: \quad \mathbf{r}=\mathbf{i}+2\mathbf{j}+\lambda(\mathbf{j}+\mathbf{k})$$

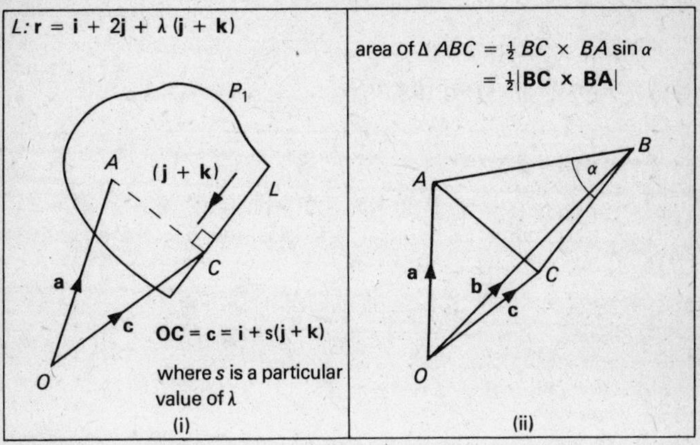

Figure 3

Refer to figure 3 (i).

Let $\mathbf{c}$ be the position vector relative to the origin of the point C on L such that $\mathbf{AC}$ is perpendicular to L.

$$\mathbf{c}=\mathbf{i}+2\mathbf{j}+s(\mathbf{j}+\mathbf{k})$$

where s is a particular value of λ in the general equation for L.

If $\mathbf{AC}$ is perpendicular to L then the scalar product of $\mathbf{AC}$ and a direction vector for L will be zero.

A direction vector for L is $\mathbf{j}+\mathbf{k}$, and
$\mathbf{AC}=(\mathbf{c}-\mathbf{a})=\mathbf{i}+2\mathbf{j}+s(\mathbf{j}+\mathbf{k})-(2\mathbf{i}-\mathbf{k})=-\mathbf{i}+(2+s)\mathbf{j}+(s+1)\mathbf{k}$.

Thus, $\mathbf{AC}.(\mathbf{j}+\mathbf{k})=3+2s=0$, and $s=-\frac{3}{2}$

Hence,

$$\mathbf{c}=\mathbf{i}+\frac{\mathbf{j}}{2}-\frac{3\mathbf{k}}{2}$$

Refer to figure 3 (ii).

Area of triangle $ABC=\frac{1}{2}|\mathbf{BC}\times\mathbf{BA}|$ and $\mathbf{BC}=\mathbf{c}-\mathbf{b}=-\frac{3}{2}(\mathbf{j}+\mathbf{k})$

$\mathbf{BA}=\mathbf{a}-\mathbf{b}=\mathbf{i}-2\mathbf{j}-\mathbf{k}$.

Therefore,

$$|\mathbf{BC}\times\mathbf{BA}|=\begin{vmatrix}\mathbf{i} & \mathbf{j} & \mathbf{k}\\ 0 & -\frac{3}{2} & -\frac{3}{2}\\ 1 & -2 & -1\end{vmatrix}=\left|-\tfrac{3}{2}\mathbf{i}+\tfrac{3}{2}\mathbf{j}+\tfrac{3}{2}\mathbf{k}\right|$$

and **the area of triangle $ABC=\dfrac{3\sqrt{3}}{4}$ units of area.**

Example 2

Two ships A and B move with constant velocities and have the following velocity and position vectors at time $t=0$:

	Velocity Vector	Position Vector
Ship A	$-\mathbf{i}+3\mathbf{j}$	$5\mathbf{i}+\mathbf{j}$
Ship B	$2\mathbf{i}+5\mathbf{j}$	$3\mathbf{i}-3\mathbf{j}$

Show that any point on the line $\mathbf{r}=4\mathbf{i}-\mathbf{j}+\lambda(2\mathbf{i}-\mathbf{j})$, where λ is a scalar, is equidistant from the ships A and B at time $t=0$.

Find the position vector of the point which is equidistant from both ships at time $t=0$ and from the point with the position vector $\mathbf{i}-\mathbf{j}$.

Calculate the value of t when the distance between A and B is least.

(A.E.B.)

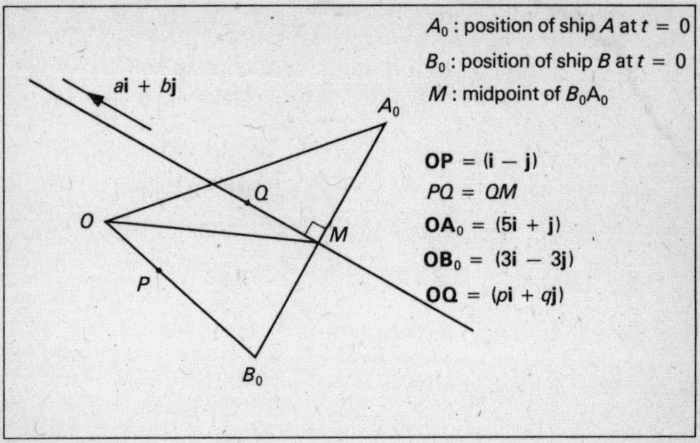

Figure 4

Refer to figure 4.

Any point on the **perpendicular bisector** of the line segment B_0A_0 will be equidistant from points A_0 and B_0.

$$\mathbf{B_0A_0} = -(3\mathbf{i} - 3\mathbf{j}) + (5\mathbf{i} + \mathbf{j}) = 2\mathbf{i} + 4\mathbf{j}$$

Let the **direction vector** for the perpendicular bisector of this line be $a\mathbf{i} + b\mathbf{j}$.

Thus,

$$(2\mathbf{i} + 4\mathbf{j}) \cdot (a\mathbf{i} + b\mathbf{j}) = 0 \quad \text{and} \quad \boldsymbol{a} = -2\boldsymbol{b}$$

The midpoint, M, of $\mathbf{B_0A_0}$ (which also lies on the perpendicular bisector of $\mathbf{B_0A_0}$), is given by

$$\mathbf{OM} = \tfrac{1}{2}(3\mathbf{i} - 3\mathbf{j} + 5\mathbf{i} + \mathbf{j}) = 4\mathbf{i} - \mathbf{j}$$

We now know the position vector of a point, M, on the perpendicular bisector of $\mathbf{B_0A_0}$ and the direction ratios for the direction vector. Thus we may now write down the vector equation of the line.

$$\mathbf{r} = 4\mathbf{i} - \mathbf{j} + \lambda(2\mathbf{i} - \mathbf{j})$$

where λ is a scalar.

Let the point, Q, which is equidistant from the ships at $t = 0$ and from the point with position vector $\mathbf{i} - \mathbf{j}$, have position vector $p\mathbf{i} + q\mathbf{j}$. We know that $(p\mathbf{i} + q\mathbf{j})$ must lie on

$$\mathbf{r} = 4\mathbf{i} - \mathbf{j} + \lambda(2\mathbf{i} - \mathbf{j})$$

Thus, $4+2\lambda=p$ (i), and $-1-\lambda=q$ (ii).

Also, since the points with position vectors $(\mathbf{i}-\mathbf{j})$ and $(3\mathbf{i}-3\mathbf{j})$ are equidistant from the point with position vector $(p\mathbf{i}+q\mathbf{j})$, then

$$|(p\mathbf{i}+q\mathbf{j})-(3\mathbf{i}-3\mathbf{j})|^2=|(p\mathbf{i}+q\mathbf{j})-(\mathbf{i}-\mathbf{j})|^2$$
$$p^2-6p+9+q^2+6q+9=p^2-2p+1+q^2+2q+1$$
$$-4p+4q+16=0$$
$$q-p+4=0.$$

Combining this result with equations (i) and (ii),

$$-1-\lambda-4-2\lambda+4=0$$
$$\lambda=-\tfrac{1}{3},$$

and the position vector $(p\mathbf{i}+q\mathbf{j})$ is given by

$$\mathbf{r}=(4\mathbf{i}-\mathbf{j})+(-\tfrac{1}{3})(2\mathbf{i}-\mathbf{j})=\frac{10\mathbf{i}}{3}-\frac{2\mathbf{j}}{3}$$

$\left(\text{i.e. } \boldsymbol{p}=\dfrac{\mathbf{10}}{\mathbf{3}}, \boldsymbol{q}=\dfrac{\mathbf{-2}}{\mathbf{3}}\right)$. Thus, $\mathbf{OQ}=\dfrac{\mathbf{10i}}{\mathbf{3}}-\dfrac{\mathbf{2j}}{\mathbf{3}}$

Refer to figure 5.

The distance between A and B at time t is given by

$$\begin{aligned}|\mathbf{B}_t\mathbf{A}_t| &= |\mathbf{r}_\mathbf{A}-\mathbf{r}_\mathbf{B}|\\ &= |(5\mathbf{i}+\mathbf{j}+t(-\mathbf{i}+3\mathbf{j}))-(3\mathbf{i}-3\mathbf{j}+t(2\mathbf{i}+5\mathbf{j}))|\\ &= |(2-3t)\mathbf{i}+(4-2t)\mathbf{j}|\\ &= ((2-3t)^2+(4-2t)^2)^{\frac{1}{2}}\end{aligned}$$

$|\mathbf{r}_\mathbf{A}-\mathbf{r}_\mathbf{B}|$ is least when $|\mathbf{r}_\mathbf{A}-\mathbf{r}_\mathbf{B}|^2$ is least.

Differentiating,

$$\frac{d}{dt}(|\mathbf{r}_\mathbf{A}-\mathbf{r}_\mathbf{B}|^2)=-6(2-3t)-4(4-2t)$$

and $\dfrac{d^2}{dt^2}(|\mathbf{r}_\mathbf{A}-\mathbf{r}_\mathbf{B}|^2)=18+8>0.$

We see that $\frac{d}{dt}(|\mathbf{r}_A - \mathbf{r}_B|^2) = 0$

when $t = \frac{14}{13}$ s.

This value of t gives **minimum** distance, $|(\mathbf{r}_A - \mathbf{r}_B)|$, between A and B, since the second differential coefficient is positive.

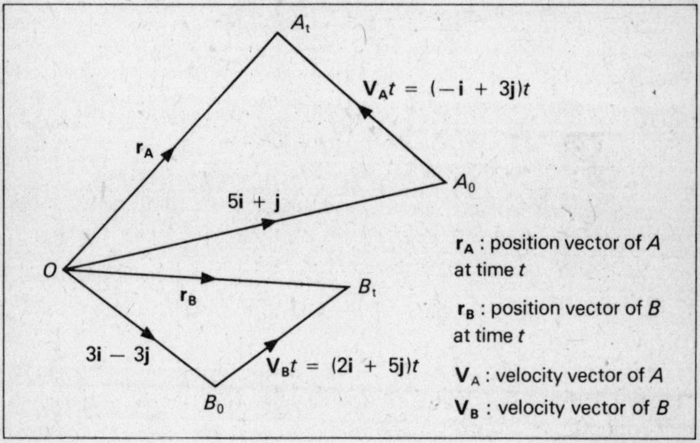

Figure 5

Example 3

The position vectors, relative to the origin, O, of four non-collinear points A, B, C, D, are $\mathbf{a}$, $\mathbf{b}$, $\frac{1}{3}\mathbf{a}$, $\frac{2}{3}\mathbf{b}$ respectively. Show that the position vector of any point on BC may be given parametrically by $\frac{1}{3}\lambda\mathbf{a} + (1-\lambda)\mathbf{b}$ and find the position vector of any point on AD. Find the position vector of the point of intersection, X, of BC and AD.

Straight lines are drawn through the midpoints of OA, OB AB, parallel to CB, DA, XO respectively. Show that these lines are concurrent and find the position vector of their point of intersection, Y.

If $OA = 15$ cm, $OB = 16$ cm and $\angle AOB = 90°$, find the length XY.

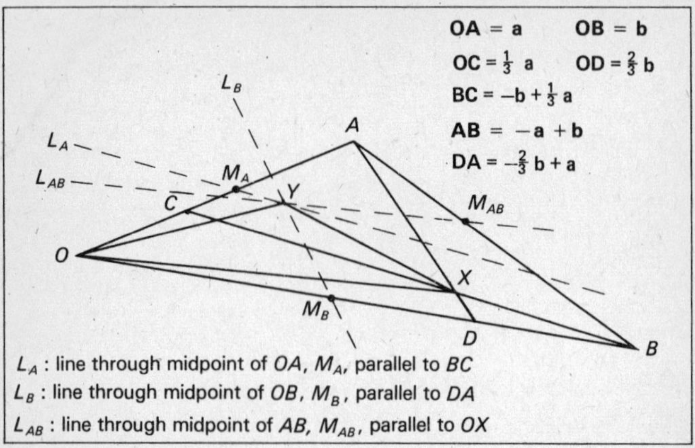

Figure 6

Refer to figure 6.

In order to give the position vector, **r**, of a point on the line *BC*, we need a direction vector, **a**, for the line and a position vector, **p**, of a particular point on the line.

The position vector may then be given parametrically in the form,

$$\text{‘}\mathbf{r}=\mathbf{p}+t\mathbf{q}\text{’}$$

where t is a constant.

Thus for *BC*, a direction vector is

$$\mathbf{BC}=\mathbf{BO}+\mathbf{OC}=-\mathbf{b}+\tfrac{1}{3}\mathbf{a}$$

Also $\mathbf{OB}=\mathbf{b}$ is the position vector of a particular point, *B*, on the line.

A position vector, **r**, of any point on the line *BC* is given by

$$\mathbf{r}=\mathbf{b}+\lambda(-\mathbf{b}+\tfrac{1}{3}\mathbf{a})=\tfrac{1}{3}\lambda\,\mathbf{a}+(1-\lambda)\mathbf{b}$$

where λ is a constant.

A direction vector for AD is

$$\mathbf{AD}=\mathbf{AO}+\mathbf{OD}=-\mathbf{a}+\tfrac{2}{3}\mathbf{b}$$

and the position vector of a particular point, A, on the line is $\mathbf{OA}=\mathbf{a}$.

A position vector, $\mathbf{s}$, of any point on the line AD is given by

$$\mathbf{s}=\mathbf{a}+\mu(-\mathbf{a}+\tfrac{2}{3}\mathbf{b})=(1-\mu)\mathbf{a}+\tfrac{2}{3}\mu\mathbf{b}$$

where μ is a constant.

If BC and AD intersect, there will be some values of μ and λ such that $\mathbf{r}=\mathbf{s}$, and these will give the position vector of the point of intersection.

Equating,

$$(1-\mu)\mathbf{a}+\tfrac{2}{3}\mu\mathbf{b}=\frac{\lambda}{3}\mathbf{a}+(1-\lambda)\mathbf{b}$$

Comparing coefficients of $\mathbf{a}$,

$$1-\mu=\frac{\lambda}{3}$$

Comparing coefficients of $\mathbf{b}$,

$$\frac{2}{3}\mu=(1-\lambda).$$

Therefore $\lambda=\dfrac{3}{7}$ and $\mu=\dfrac{6}{7}$

and **the position vector of the point of intersection, *X*, of *AD* and *BC* is**

$$\frac{1}{7}(\mathbf{a}+4\mathbf{b}).$$

We shall now find the parametric equations for the position vectors $\mathbf{r}_\mathbf{A}$, $\mathbf{r}_\mathbf{B}$, $\mathbf{r}_\mathbf{AB}$ of points on the lines L_A, L_B, L_{AB}, respectively.

L_A: A direction vector is **BC**, and the position vector of a particular point on the line is $\mathbf{OM_A}$.

Therefore $\mathbf{r_A}=\frac{1}{2}\mathbf{a}+\alpha\left(-\mathbf{b}+\frac{1}{3}\mathbf{a}\right)=\left(\frac{1}{2}+\frac{\alpha}{3}\right)\mathbf{a}-\alpha\mathbf{b}$

where α is a constant.

L_B: A direction vector is **DA** and the position vector of a particular point on the line is $\mathbf{OM_B}$.

Therefore $\mathbf{r_B}=\frac{1}{2}\mathbf{b}+\beta\left(-\frac{2}{3}\mathbf{b}+\mathbf{a}\right)=\beta\mathbf{a}+\left(\frac{1}{2}-\frac{2}{3}\beta\right)\mathbf{b}$

where β is a constant.

L_{AB}: A direction vector is **OX** where X is the point of intersection of BC and AD. The position vector of a particular point on the line is $\mathbf{OM_{AB}}$.
Therefore,

$$\mathbf{r_{AB}}=\frac{1}{2}(\mathbf{a}+\mathbf{b})+\gamma(\mathbf{a}+4\mathbf{b})=\left(\frac{1}{2}+\gamma\right)\mathbf{a}+\left(\frac{1}{2}+4\gamma\right)\mathbf{b}$$

where γ is a constant.

If the lines L_A, L_B, L_{AB}, are concurrent then there are values of α, β, γ, such that $\mathbf{r_A}=\mathbf{r_B}=\mathbf{r_{AB}}$.

Equating $\mathbf{r_A}$ and $\mathbf{r_B}$,

$$\left(\frac{1}{2}+\frac{\alpha}{3}\right)\mathbf{a}-\alpha\mathbf{b}=\beta\mathbf{a}+\left(\frac{1}{2}-\frac{2}{3}\beta\right)\mathbf{b}$$

Comparing coefficients of **a** and **b**,

$$\frac{1}{2}+\frac{\alpha}{3}=\beta \quad \text{and} \quad -\alpha=\frac{1}{2}-\frac{2}{3}\beta.$$

Therefore $\boldsymbol{\beta}=\frac{\mathbf{3}}{\mathbf{7}}$ and $\boldsymbol{\alpha}=-\frac{\mathbf{3}}{\mathbf{14}}$

Equating $\mathbf{r_B}$ and $\mathbf{r_{AB}}$,

$$\beta\mathbf{a}+\left(\frac{1}{2}-\frac{2}{3}\beta\right)\mathbf{b}=\left(\frac{1}{2}+\gamma\right)\mathbf{a}+\left(\frac{1}{2}+4\gamma\right)\mathbf{b}$$

Comparing coefficients of **a** and **b**

$$\frac{1}{2}+\gamma=\beta \quad \text{and} \quad \frac{1}{2}+4\gamma=\frac{1}{2}-\frac{2}{3}\beta$$

Therefore $\boldsymbol{\beta}=\frac{\mathbf{3}}{\mathbf{7}}$ and $\gamma=-\frac{\mathbf{1}}{\mathbf{14}}$.

These values for α, β and γ are consistent with concurrence and **the position vector of the point of intersection, *Y*, is**

$$\frac{3}{7}\mathbf{a}+\frac{3}{14}\mathbf{b}$$

We are given that $OA=15\,\text{cm}$, $OB=16\,\text{cm}$, and $\angle AOB=90°$, and we are asked to find the length of **XY**.

$$\mathbf{XY}=\mathbf{XO}+\mathbf{OY}=-\frac{1}{7}\mathbf{a}-\frac{4}{7}\mathbf{b}+\frac{3}{7}\mathbf{a}+\frac{3}{14}\mathbf{b}=\frac{2}{7}\mathbf{a}-\frac{5}{14}\mathbf{b}.$$

In order to find the length of **XY** we shall consider the scalar product of **XY** with itself.

$$\mathbf{XY}.\mathbf{XY}=|\mathbf{XY}|\,|\mathbf{XY}|\cos 0°=|\mathbf{XY}|^2$$

Since OA is perpendicular to OB then **XY** may be expressed in terms of unit perpendicular vectors $\hat{\mathbf{a}}$ and $\hat{\mathbf{b}}$.

Thus

$$\mathbf{XY}=\frac{2}{7}\times 15\,\hat{\mathbf{a}}-\frac{5}{14}\times 16\hat{\mathbf{b}} \text{ and}$$

$$\mathbf{XY}.\mathbf{XY}=|\mathbf{XY}|^2=\left(\frac{30}{7}\right)^2+\left(\frac{40}{7}\right)^2.$$

Therefore

$$|\mathbf{XY}|=\frac{\mathbf{50}}{\mathbf{7}}$$

Chapter 2
Motion of Particles Moving in Straight Lines

Example 1

Two particles of masses $3m$ and m are connected by a light inextensible string, passing over a smooth fixed light pulley; the string is taut and the straight parts vertical when the system is released from rest. Show that the acceleration of either particle is $g/2$ and that the tension in the string is $3mg/2$.

The particles start at the same level. The string breaks after time t_1 and the string is sufficiently long so that the lighter particle does not hit the pulley. Find the vertical distance between the particles when the lighter one reaches its greatest height.

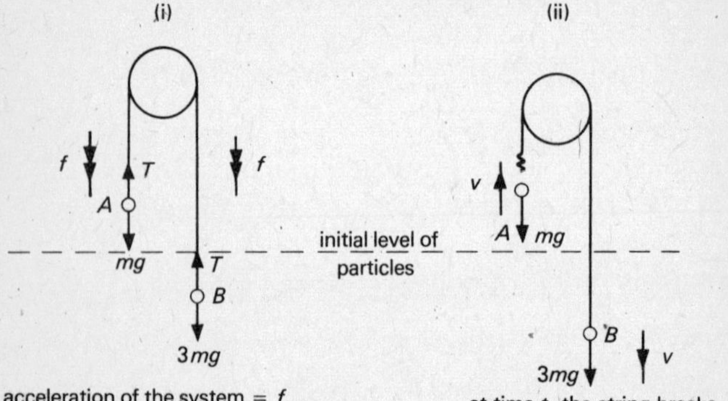

acceleration of the system = f
tension in the string = T

at time t_1 the string breaks,
the particles have velocity v

Figure 7

Refer to figure 7 (i).

Since the pulley is smooth and light, and the string light, the tension T will be transmitted throughout the string.

Applying Newton's second law, '$f=ma$', to the system,

$$\text{particle } A\text{: } T-mg=mf$$

$$\text{particle } B\text{: } 3mg-T=3mf.$$

Combining these equations gives $\boldsymbol{f=\frac{g}{2}}$ and $\boldsymbol{T=\frac{3mg}{2}}$.

The particles move with this constant acceleration, $g/2$, for the time t, and then the string breaks. Applying the equations of motion for constant acceleration:

$$\text{'}v=u+at\text{'},\quad v=\frac{gt_1}{2} \quad \text{and} \quad \text{'}s=ut+\tfrac{1}{2}at^2\text{'},\quad s=\frac{gt_1^2}{4}$$

where v is the speed of the particles when the string breaks and s is the distance travelled by each particle before the string breaks.

Refer to figure 7 (ii).

The string now breaks and the particles move freely under gravity. Applying the equations of motion for constant acceleration to this part of the motion, until A comes instantaneously to rest at the highest point of the motion, where t and s_1 are the time taken and distance travelled respectively:

$$\text{'}v=u+at\text{'},\quad 0=\frac{gt_1}{2}-gt,\quad \text{i.e. } \boldsymbol{t=\frac{t_1}{2}}$$

$$\text{and '}s=vt-\tfrac{1}{2}at^2\text{'},\quad \boldsymbol{s_1=\frac{gt_1^2}{8}}.$$

Let s_2 be the distance that B travels in time $t_1/2$ after the string breaks.

$$\text{'}s=ut+\tfrac{1}{2}at^2\text{'},$$

$$s_2=\frac{g}{2}t_1\frac{t_1}{2}+\tfrac{1}{2}\frac{gt_1^2}{4},\quad \text{i.e. } s_2=\frac{3gt_1^2}{8}.$$

Hence **the total distance between the particles at this instant** is (distance A has travelled upwards)+(distance B has travelled downwards)

$$=\left(\frac{gt_1^2}{4}+\frac{gt_1^2}{8}\right)+\left(\frac{gt_1^2}{4}+\frac{3gt_1^2}{8}\right)=\boldsymbol{gt_1^2}.$$

Example 2

A car, A, accelerates from rest up to its maximum speed, V, with constant acceleration, a. At the instant it starts it is overtaken by car, B, moving with constant speed, U. A catches up with B after distance S (where $S > V^2/2a$). Find the time taken by A. Derive a quadratic equation in V with coefficients in terms of U, S, a and hence find V.

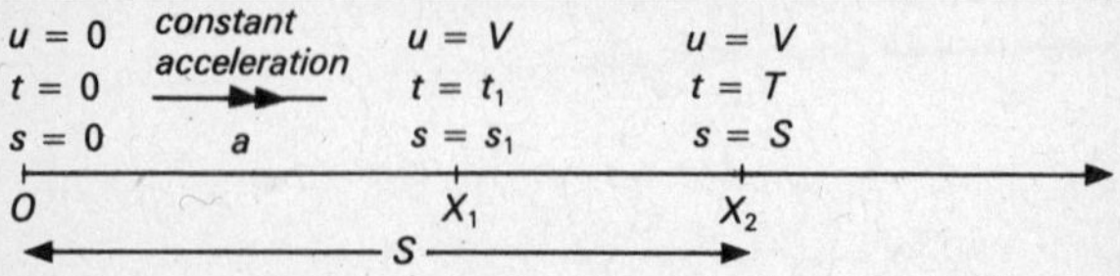

let X_1 be the point at which A achieves its maximum speed V, and X_2 be the point at which A overtakes B

Figure 8

Refer to figure 8.

Applying the equations of motion for constant acceleration to the journey OX_1 for A:

$$\text{'}v = u + at\text{'}, \quad V = 0 + at_1, \quad t_1 = \frac{V}{a}$$

$$\text{'}s = ut + \tfrac{1}{2}at^2\text{'}, \quad s_1 = 0 + \frac{a}{2}\frac{V^2}{a^2}, \quad s_1 = \frac{V^2}{2a}.$$

Applying the equations of motion for **constant speed** for the journey X_1X_2 for A:

'distance = speed × time', $S - s_1 = V(T - t_1)$.

Considering the motion of B for OX_2:

$$S = UT, \quad T = \frac{S}{U}.$$

Combining these equations,

$$S - \frac{V^2}{2a} = V\left(\frac{S}{U} - \frac{V}{a}\right), \quad V^2U - 2aVS + 2aUS = 0.$$

Hence

$$V = \frac{2aS \pm (4a^2S^2 - 4U^2 2aS)^{\frac{1}{2}}}{2U}$$

$$= \frac{aS}{U}\left(1 \pm \left(1 - \frac{2U^2}{aS}\right)^{\frac{1}{2}}\right).$$

We have to choose between the two possible roots. We are given that

$$S > \frac{V^2}{2a}, \quad \text{i.e. } S > s_1.$$

Hence we know that the time for which A is travelling at constant speed, $T - t_1$, is greater than 0.

Thus,

$$\frac{S}{U} > \frac{V}{a}, \quad V < \frac{aS}{U}$$

and we must take the smaller root of the quadratic equation.

Therefore,

$$V = \frac{aS}{U}\left(1 - \left(1 - \frac{2U^2}{aS}\right)^{\frac{1}{2}}\right).$$

Example 3

(a) A train starts from a station accelerating uniformly. It reaches a certain speed and immediately decelerates uniformly, coming to rest at the next station.

Prove that the train's average speed is half of its maximum speed.

(b) A particle is projected in a straight line with a certain velocity and a constant acceleration. One second later, another particle is projected after it from the same point with half the velocity and double acceleration. When it overtakes the first, the velocities of the two particles are, respectively, 22 and 31 ms^{-1}. Prove that both have travelled 48 m.

(S.U.J.B.)

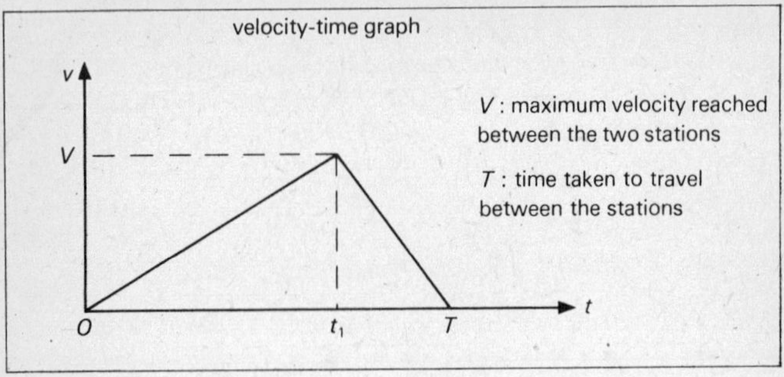

Figure 9

(a) Refer to figure 9.

The **area under the velocity-time graph represents the total distance travelled, *S*,** between the stations.

Thus,

$$S=\frac{1}{2}t_1V+\frac{1}{2}(T-t_1)V=\frac{1}{2}VT.$$

The **average speed of the train**

$$=\frac{\textbf{total distance between the stations}}{\textbf{total time taken}}.$$

$$=\frac{S}{T}=\frac{V}{2}$$

Therefore **the average speed of the train is half of its maximum speed.**

(b)

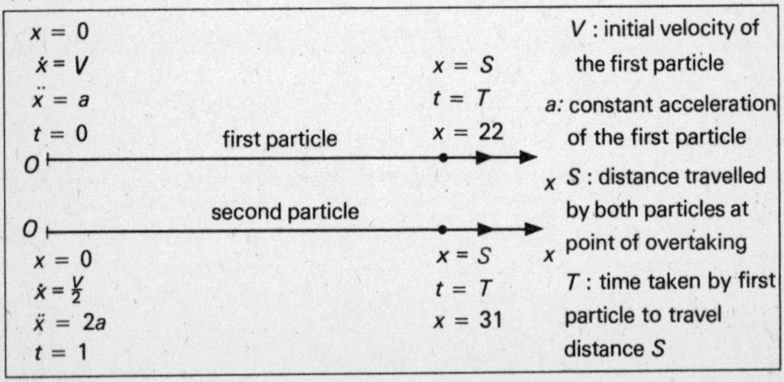

Figure 10

Refer to figure 10.

Both particles move with constant acceleration. We shall, therefore, apply **the standard equations of motion for constant acceleration** to both particles for the journey from $x=0$ to $x=S$.

'$v=u^2+2as$'

1st particle: $22\times 22=V^2+2aS$

2nd particle: $31\times 31=\frac{V^2}{4}+16aS$.

Eliminating V^2,

$$\boldsymbol{a=\frac{240}{S}} \qquad \text{(i)}$$

'$v=u+at$'

1st particle: $22=V+aT$ (ii)

2nd particle: $31=\frac{V}{2}+2a(T-1)$. (iii)

Eliminating V,

$$\boldsymbol{T=\frac{40+4a}{3a}}. \qquad \text{(iv)}$$

Substituting from equation (i),

$$\boldsymbol{T=\frac{S+24}{18}} \qquad \text{(v)}$$

'$s=vt-\frac{1}{2}at^2$'

1st particle: $S=22T-\frac{1}{2}aT^2$

Substituting from equations (i) and (v),

$$S=22\left(\frac{S+24}{18}\right)-\frac{1}{2}\,\frac{240}{S}\left(\frac{S+24}{18}\right)^2.$$

Simplifying, $S^2-78\,S+1440=0$

$$S=\frac{78\pm 18}{2}$$

$$\boldsymbol{S=30\,\text{m} \text{ or } S=48\,\text{m}}.$$

Substituting $\boldsymbol{S}=\mathbf{30}$, back into equations (v) and (i) gives $T=3$ and $a=8$.

From equation (ii), $V=-2$ **which is impossible and we reject this value of** S.

Substituting $\boldsymbol{S}=\mathbf{48}$ into equations (v) and (i) gives $T=4$, $a=5$.

From equation (ii), $V=2$. Checking from equation (iii), $V=\mathbf{2}$ also.

Therefore the second particle overtakes the 1st particle when both particles have travelled 48 m.

Chapter 3
Friction

Example 1

Explain, with the aid of a diagram, the meaning of the expression **coefficient of friction.**

A straight uniform rod of AB of weight W rests in limiting equilibrium with the end A on horizontal ground and the end B against a vertical wall. The vertical plane containing AB is perpendicular to the wall. The coefficient of friction between the rod and the ground is 4/5. The coefficient of friction between the rod and the wall is 3/5. Given that the inclination of AB to the horizontal is α, calculate:

(i) the normal reactions at A and B in terms of W.
(ii) the numerical value of $\tan \alpha$.

(A.E.B.)

μ : coefficient of friction

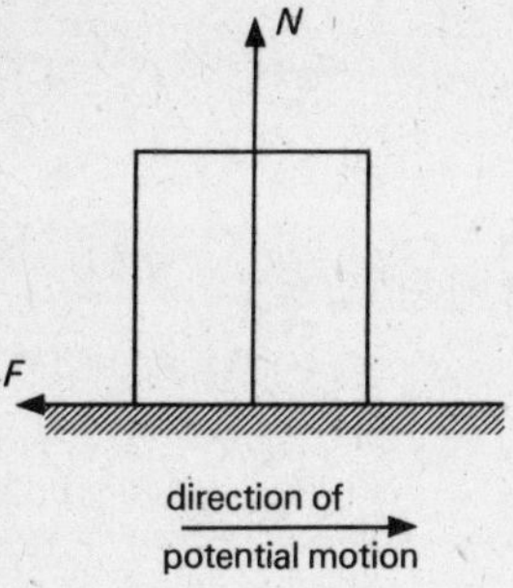

when the frictional force, F, reaches its **limiting** value then $\mu = \frac{F}{N}$

Figure 11

Refer to figure 11.

Friction is the name given to the force which comes into play between two surfaces in **rough** contact in order to resist potential and actual motion between them. Up to a **limiting** value, the magnitude of the frictional force is just sufficient to prevent motion. The limiting value of the frictional force is directly proportional to the normal reaction between the surfaces, the constant of proportionality being the **coefficient of friction.** The coefficient of friction is determined by the nature of the surfaces in contact.

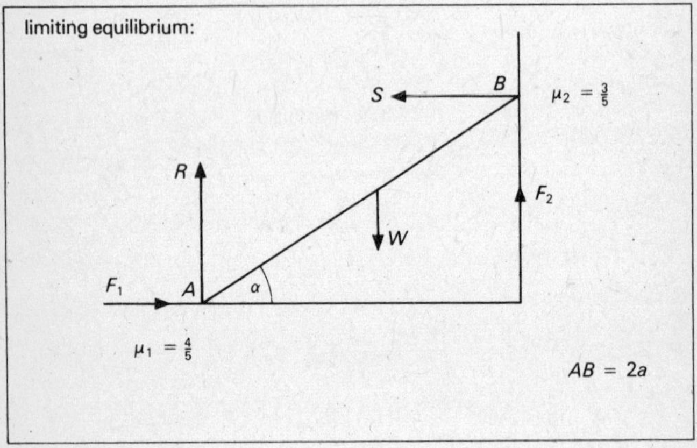

Figure 12

Refer to figure 12.

Applying the conditions for equilibrium:

Resolving horizontally,

$$F_1 = S$$

Resolving vertically,

$$R + F_2 = W$$

Taking moments about A,

$$-a\cos\alpha\, W + 2a\cos\alpha\, F_2 + 2a\sin\alpha\, S = 0.$$

We are given that the equilibrium is **limiting**, thus

$$F_1=\frac{4}{5}R \text{ and } F_2=\frac{3}{5}S$$

Combining these equations,

(i) $$S=\frac{4}{5}R, \quad R+\frac{3}{5}S=W$$

$$\boldsymbol{R=\frac{25}{37}W, \quad S=\frac{20}{37}W}$$

(ii) $$-a\cos\alpha\, W+2a\cos\alpha\frac{3}{5}\times\frac{20}{37}W+2a\sin\alpha\frac{20}{37}W=0,$$

$$\boldsymbol{\tan\alpha=\frac{13}{40}}.$$

Example 2

Using a diagram, explain the meaning of the term, *angle of friction.*

A uniform straight pole AB, of length 2.5 metres, stands in limiting equilibrium with the end A on horizontal ground and the end B against a vertical wall. The vertical plane containing AB is perpendicular to the wall and A is a distance 0.7 metres from the wall. The angles of friction between the pole and the ground and between the pole and the wall are equal. Show that the line of action of the resultant force exerted on the pole by the ground at A is perpendicular to the line of action of the resultant force exerted on the pole by the wall at B.

Hence, or otherwise, show that the coefficient of friction at A and B is 1/7. Given that the pole weighs 50 N, find the magnitude of the resultant reactions:
(i) between the pole and the ground
(ii) between the pole and the wall.

(A.E.B.)

Refer to figure 13 for the explanation of the term, angle of friction.

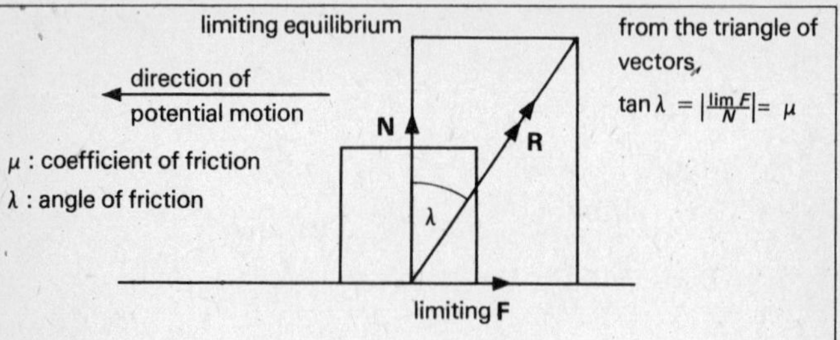

the **angle of friction,** λ, is the angle between the **normal** reaction, **N**, and the **resultant** reaction, **R**, when **friction** is limiting

Figure 13

Refer to figure 14.

AC is the line of action of the resultant reaction, R, at A.
BC is the line of action of the resultant reaction, S, at B.

The system is in equilibrium (limiting) under the action of three non-parallel, coplanar forces R, S and W. Thus, the lines of action of the forces are **concurrent,** meeting at C.

$$\angle CAB = 90^\circ - (\theta + \lambda)$$

and

$$\angle CBA = (\theta + \lambda)$$

Therefore $\angle \boldsymbol{ACB} = 180^\circ - (90^\circ - (\theta + \lambda) + (\theta + \lambda))$

$$= \mathbf{90^\circ}$$

The coefficient of friction, $\mu = \tan\lambda$.

In $\triangle ACX$,

$$\tan\lambda = \frac{CX}{AX} = \frac{CX}{AX + YX}$$

but

$$XY = CZ = BZ\tan\lambda = 0.35\tan\lambda$$

Therefore,

$$\tan\lambda = \frac{0.35}{2.4 + 0.35\tan\lambda}$$

and $$0.35 \tan^2\lambda + 2.4 \tan\lambda - 0.35 = 0.$$

Multiply through by 20,

$$7 \tan^2 \lambda + 48 \tan\lambda - 7 = 0$$
$$(7 \tan\lambda - 1)(\tan\lambda + 7) = 0$$

either $$\tan\lambda = \frac{1}{7} \text{ or } \tan\lambda = -7.$$

But $0° \leqslant \lambda \leqslant 90°$ and we reject the negative value for $\tan\lambda$.

Therefore $$\boldsymbol{\mu = \tan\lambda = \frac{1}{7}}.$$

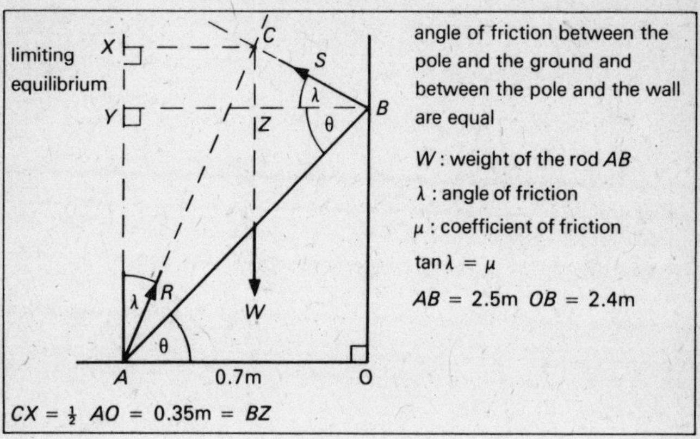

Figure 14

We are given that $W = 50$ N.

Resolving vertically,

$$R \cos\lambda + S \sin\lambda = W$$

and horizontally,

$$R \sin\lambda = S \cos\lambda$$
$$R = 7S$$

Substituting,

$$7S + S \tan \lambda = 50 \sec \lambda$$

$$\frac{50}{7} S = \frac{50 \times \sqrt{50}}{7}$$

$$\boldsymbol{S = 5\sqrt{2}} \quad \text{(ii)}$$

and $$\boldsymbol{R = 35\sqrt{2}}. \quad \text{(i)}$$

Chapter 4
Relative Motion

Example 1

When a motorist is driving with velocity $3\mathbf{i}+4\mathbf{j}$, the wind appears to come from the direction $\mathbf{i}$. When he doubles his velocity, the wind appears to come from the direction $\mathbf{i}+\mathbf{j}$. Prove that the true velocity of the wind is $2\mathbf{i}+4\mathbf{j}$. The motorist changes his speed but still drives in the same direction. If the wind appears to come from the direction $\mathbf{i}+\frac{1}{2}\mathbf{j}$, calculate the motorist's speed.

(A.E.B.)

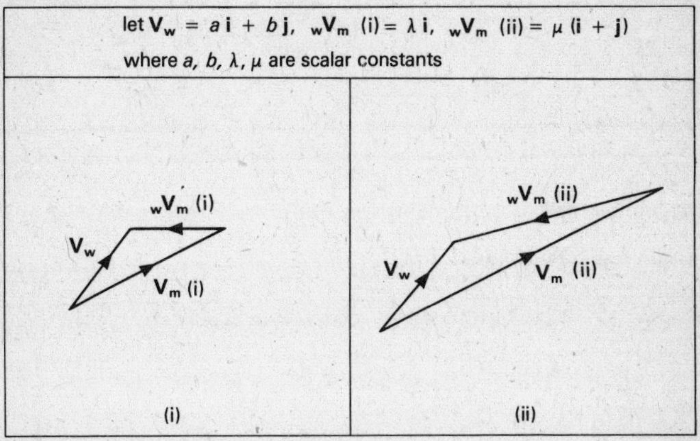

Figure 15

Refer to figure 15(i).

The initial velocity of the motorist, $\mathbf{V}_m$(i), is $3\mathbf{i}+4\mathbf{j}$. Let the velocity of the wind $\mathbf{V}_w = a\mathbf{i}+b\mathbf{j}$, and the velocity of the wind relative to the motorist initially ${}_w\mathbf{V}_m(\mathrm{i}) = \lambda\mathbf{i}$.

Since ${}_w\mathbf{V}_m(\mathrm{i}) = \mathbf{V}_w - \mathbf{V}_m(\mathrm{i})$, i.e. $\lambda\mathbf{i} = a\mathbf{i}+b\mathbf{j}-(3\mathbf{i}+4\mathbf{j})$, then equating coefficients of $\mathbf{i}$ and $\mathbf{j}$:

$$\lambda = a-3 \text{ and } b=4.$$

The motorist now doubles his velocity to $\mathbf{V_m}(ii) = 6\mathbf{i} + 8\mathbf{j}$ and the velocity of the wind relative to the motorist changes to ${}_{\mathbf{w}}\mathbf{V_m}(ii) = \mu(\mathbf{i} + \mathbf{j})$. (Refer to figure 15(ii)).

But ${}_{\mathbf{w}}\mathbf{V_m}(ii) = \mathbf{V_w} - \mathbf{V_m}(ii) = a\mathbf{i} + b\mathbf{j} - (6\mathbf{i} + 8\mathbf{j}) = \mu(\mathbf{i} + \mathbf{j})$.

Equating coefficients of **i** and **j**:

$a - 6 = \mu$ and $b - 8 = \mu$. Therefore $a = 2$, and the **true velocity of the wind**, $\mathbf{V_w} = 2\mathbf{i} + 4\mathbf{j}$.

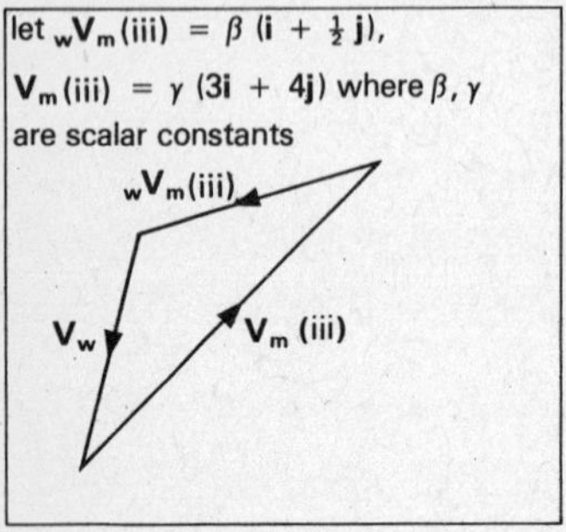

Figure 16

Let the new velocity of the motorist,

$$\mathbf{V_m}(iii) = \gamma(3\mathbf{i} + 4\mathbf{j})$$

We are given that the wind now appears to come from $\mathbf{i} + \frac{1}{2}\mathbf{j}$.

$$\text{i.e. } {}_{\mathbf{w}}\mathbf{V_m}(iii) = \mathbf{V_w} - \mathbf{V_m}(iii)$$
$$= (2\mathbf{i} + 4\mathbf{j}) - \gamma(3\mathbf{i} + 4\mathbf{j})$$
$$= \beta(\mathbf{i} + \tfrac{1}{2}\mathbf{j}).$$

Where γ and β are scalar constants.

Therefore, $(2 - 3\gamma)\mathbf{i} + (4 - 4\gamma)\mathbf{j} = \beta\mathbf{i} + \frac{1}{2}\beta\mathbf{j}$.

Equating coefficients of **i** and **j**:

$$2 - 3\gamma = \beta \text{ and } 4 - 4\gamma = \tfrac{1}{2}\beta, \text{ hence } \gamma = \frac{6}{5}.$$

Therefore $\mathbf{V_m}(iii) = \frac{6}{5}(3\mathbf{i} + 4\mathbf{j})$ and $|\mathbf{V_m}(iii)| = 6$ units.

Thus **the motorist's speed is now 6 units.**

Example 2

Ship P is moving with constant velocity $\mathbf{v_P}$ and ship Q is moving with constant velocity $\mathbf{v_Q}$. At time $t = 0$ the position vector of P relative to Q is **a**.
(i) Find the position vector of P relative to Q at some later time t.

(ii) Show that the ships are at their minimum distance apart at time

$$t = -\frac{\mathbf{a}.(\mathbf{v_P}-\mathbf{v_Q})}{(\mathbf{v_P}-\mathbf{v_Q}).(\mathbf{v_P}-\mathbf{v_Q})}.$$

Given that P is travelling in a north-westerly direction at $5\sqrt{2}\,\text{ms}^{-1}$, and Q is travelling due north at $10\,\text{ms}^{-1}$, and that at 12.00 hours P is 3 km due west of Q, find the minimum distance apart of the ships and the time at which this occurs.

(W.J.E.C.)

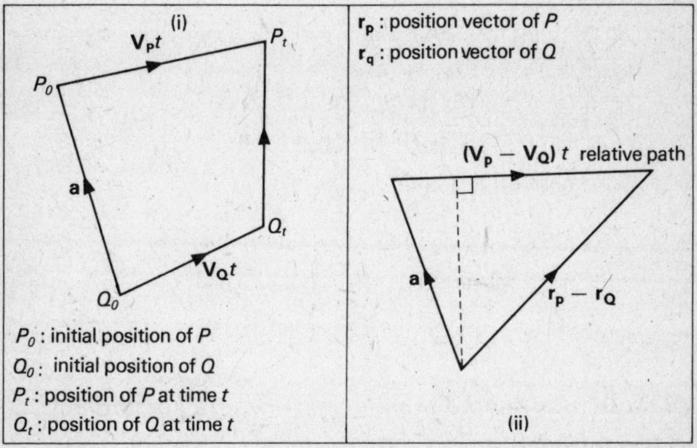

Figure 17

Refer to figure 17(i).

From the diagram we see that the position vector of P relative to Q at time t, $(\mathbf{r_P}-\mathbf{r_Q})$, is given by

$$(\mathbf{r_P}-\mathbf{r_Q}) = \mathbf{Q_tP_t} = -\mathbf{v_Q}t + \mathbf{a} + \mathbf{v_P}t = \mathbf{a} + (\mathbf{v_P}-\mathbf{v_Q})t$$

From the above result we draw the vector diagram in figure 17(ii).

We see that the ships are at their minimum distance apart when $(\mathbf{r_P}-\mathbf{r_Q})$ is perpendicular to the relative path $(\mathbf{v_P}-\mathbf{v_Q})t$.

i.e. when $(\mathbf{a}+(\mathbf{v}_P-\mathbf{v}_Q)t)\cdot(\mathbf{v}_P-\mathbf{v}_Q)t=0$.

Since the scalar product is distributive over addition, and $t\neq 0$.

$$\mathbf{a}\cdot(\mathbf{v}_P-\mathbf{v}_Q)+(\mathbf{v}_P-\mathbf{v}_Q)t\cdot(\mathbf{v}_P-\mathbf{v}_Q)=0,$$

$$t=\frac{-\mathbf{a}\cdot(\mathbf{v}_P-\mathbf{v}_Q)}{(\mathbf{v}_P-\mathbf{v}_Q)\cdot(\mathbf{v}_P-\mathbf{v}_Q)}.$$

For the final part of the question we shall take $\mathbf{i}$ as the unit vector in an easterly direction and $\mathbf{j}$ as the unit vector in the northerly direction. With the usual notation, and taking $t=0$ as 1200 hours.

$$\mathbf{v}_P=\frac{5\sqrt{2}}{\sqrt{2}}(-\mathbf{i}+\mathbf{j}),\quad \mathbf{v}_Q=10\mathbf{j},\quad \mathbf{a}=-3000\mathbf{i}.$$

Thus the minimum distance apart of the ships, from the above result, occurs when

$$t=-\frac{-3000\mathbf{i}\cdot(-5\mathbf{i}-5\mathbf{j})}{(-5\mathbf{i}-5\mathbf{j})\cdot(-5\mathbf{i}-5\mathbf{j})}=\frac{-15000}{50}=-\mathbf{300}\text{s}$$

i.e. at 300s before 1200 hours.

When $t=-300$,

$$\begin{aligned}|\mathbf{r}_P-\mathbf{r}_Q|&=|-3000\mathbf{i}+(-5\mathbf{i}-5\mathbf{j})\times-300|\\&=|-1500\mathbf{i}+1500\mathbf{j}|=1500\sqrt{2}\,\text{m}.\end{aligned}$$

Thus **the minimum distance apart is $\frac{3}{2}\sqrt{2}$ km and occurs at 5 minutes before 1200 hours.**

Example 3

A helicopter has a maximum speed of $108\,\text{km h}^{-1}$ in still air. The helicopter flies on a triangular circuit ABC, where $AB=30\,\text{km}$, $BC=40\,\text{km}$ and $CA=50\,\text{km}$. If the time taken to change directions can be neglected, find the least time for the helicopter to fly the circuit in the sense AB, BC, CA, when a wind of $15\,\text{km h}^{-1}$ blows in the direction CA.

(W.E.J.C.)

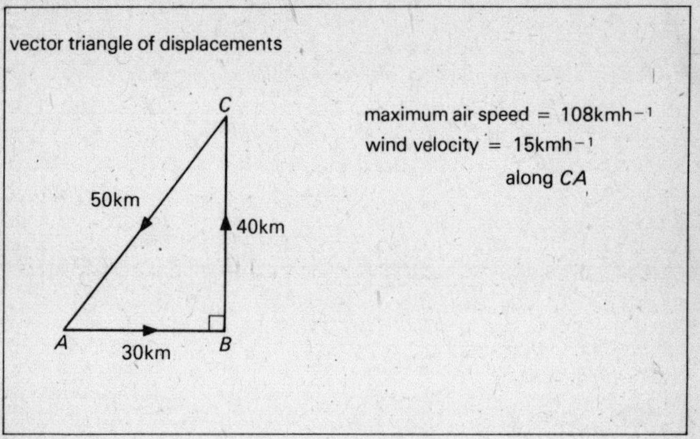

Figure 18

Refer to figure 18.

The helicopter must complete its triangular circuit *ABC* in the least possible time. For each part *AB*, *BC*, *CA*, of the journey, the resultant velocity of the helicopter is the resultant of the wind velocity, $15\,\text{km}\,\text{h}^{-1}$, in the direction *CA* and the **maximum** still air velocity, $108\,\text{km}\,\text{h}^{-1}$, of the helicopter in a direction chosen so that the **resultant** velocity is along *AB*, *BC*, *CA* in turn. We shall find the magnitudes of the resultant velocities for the parts *AB*, *BC* of the journey, by **scale drawing**.

Refer to figure 19(i).

A base line is drawn in direction *AB*. A point *B′* is marked off. A line is drawn through *B′* at angle $\arctan\frac{4}{3}$ to the base line. *D′* is marked off on this line where *B′D′* represents the wind velocity. *D′A′* is now drawn where *D′A′* represents the maximum still air speed of $108\,\text{km}\,\text{h}^{-1}$. **By measurement, the resultant velocity along *AB* has magnitude, $98\,\text{km}\,\text{h}^{-1}$.**

Refer to figure 19(ii).

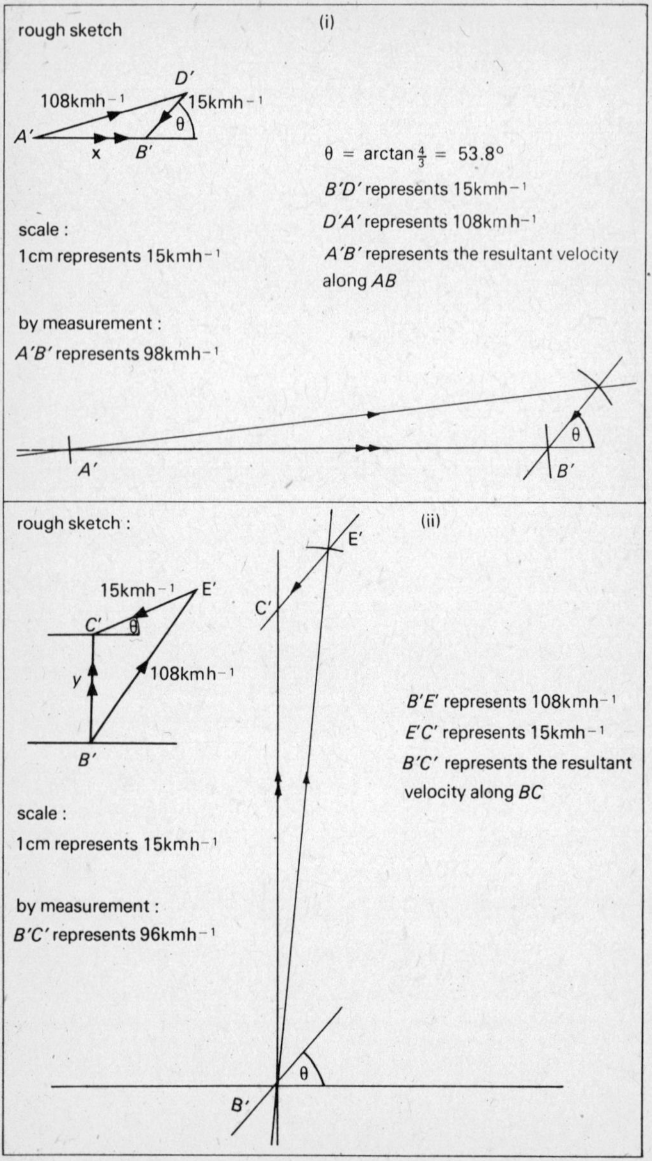
(i)
rough sketch
108kmh−1
D′
15kmh−1
θ
A′
x
B′
θ = arctan 4/3 = 53.8°
B′D′ represents 15kmh−1
D′A′ represents 108kmh−1
A′B′ represents the resultant velocity along AB
scale :
1cm represents 15kmh−1
by measurement :
A′B′ represents 98kmh−1
θ
A′
B′
(ii)
rough sketch :
15kmh−1
E′
C′
θ
y
108kmh−1
B′
E′
C′
B′E′ represents 108kmh−1
E′C′ represents 15kmh−1
B′C′ represents the resultant velocity along BC
scale :
1cm represents 15kmh−1
by measurement :
B′C′ represents 96kmh−1
θ
B′

Figure 19

A base line in the direction AB is drawn. B' is marked off. A line perpendicular to the base line is drawn through B'. A circle centre, B', radius representing 108 km h^{-1}, is drawn. A line at angle arctan $\frac{4}{3}$ to the base line is drawn through B'. A line parallel to this is drawn, such that the distance from the perpendicular through B' to the circle, represents 15 km h^{-1}.

By measurement, the resultant velocity along *BC* has magnitude 96 km h^{-1}.

(These two resultant velocities could also have been found by applying the **cosine rule** to the triangles $A'B'D'$ (figure 19(i)) and $B'C'E'$ (figure 19(ii)), respectively.

By calculation, the resultant velocity along *AB* has magnitude 98.33 km h^{-1} and along *BC* has magnitude 95.62 km h^{-1}.)

The resultant velocity along *CA* is (108 + 15) km h^{-1} **= 123 km h^{-1}**, as the wind velocity is in this direction.

Therefore from 'time = distance/speed' for **constant** velocity,

the total time, *T*, for the fastest journey, *AB*, *BC*, *CA* is given by

$$T = \frac{30}{98} + \frac{40}{98} + \frac{50}{123} = \textbf{1 hour 8 minutes.}$$

Chapter 5

General Motion of Particles in Planes

Example 1

A particle, P, moves on a parabola whose equation is $ay=x^2$, referred to orthogonal Cartesian axes Ox, Oy. The component of the velocity of P in the Ox direction is a constant, V. If the particle passes through O at time $t=0$, show that at time $t=T$, its speed will be

$$V\left(1+\frac{4V^2T^2}{a^2}\right)^{\frac{1}{2}}.$$

At $t=0$, a second particle is projected from rest at 0, with constant acceleration f along the straight line $y=mx$ $(m>0)$ referred to the same axes.

Show that the particles will collide only if $af>2V^2$, and

$$m=\frac{2V^2}{(a^2f^2-4V^4)^{\frac{1}{2}}},$$

and find the time at which the collision occurs.

(W.J.E.C.)

Refer to figure 20(i).

Consider the motion of P:

$$ay=x^2.$$

Differentiating w.r.t. time, $a\dot{y}=2x\dot{x}$.

But, $\dot{x}=V$ and $x=\sqrt{ay}$,

thus, $$\dot{y}=2\sqrt{\frac{y}{a}}V.$$

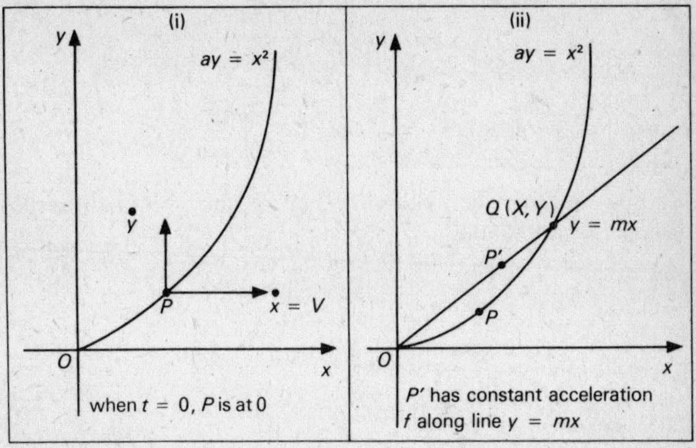

Figure 20

The speed of the particle is therefore given by

$$\sqrt{\dot{x}^2+\dot{y}^2}=\sqrt{V^2+\frac{4V^4T^2}{a^2}}=V\left(1+\frac{4V^2T^2}{a^2}\right)^{\frac{1}{2}}.$$

At time $t=T$, (from 'distance = speed × time' for constant speed), $x=VT$, and therefore

$$y=\frac{V^2T^2}{a}$$

Refer to figure 20(ii).

A second particle, P', is projected from O at time $t=0$ along the line $y=mx$, with constant acceleration f. If P and P' are to collide, then they must be at the same point, $Q(X, Y)$, at the same time, t, after projection.

Consider the motion of P' along the line $y=mx$. Using the standard equation of motion, '$s=ut+\frac{1}{2}at^2$', for constant acceleration,

$$OQ=\tfrac{1}{2}ft^2, \text{ but } OQ=\sqrt{X^2+Y^2} \text{ and } Y=mX.$$

Thus

$$(x^2+m^2X^2)^{\frac{1}{2}}=\frac{ft^2}{2}$$

$$\boldsymbol{X=\frac{ft^2}{2(1+m^2)^{\frac{1}{2}}}} \text{ and } \boldsymbol{Y=\frac{mft^2}{2(1+m^2)^{\frac{1}{2}}}}$$

We now consider the motion of P. After time t, P has travelled distance Vt horizontally

and $$\frac{V^2t^2}{a} \text{ vertically.}$$

For collision, these co-ordinates for P must be identical to those for P'.

Equating co-ordinates:

Horizontally,

$$\frac{ft^2}{2(1+m^2)^{\frac{1}{2}}}=Vt$$

thus, $t=0$, which is the initial position, or

$$\boldsymbol{t=\frac{2(1+m^2)^{\frac{1}{2}}}{f}V.}$$

Vertically,

$$\frac{mft^2}{2(1+m^2)^{\frac{1}{2}}}=\frac{V^2t^2}{a}$$

thus, $$m=\frac{2V^2}{(a^2f^2-4V^4)^{\frac{1}{2}}}$$

which has real values for $(a^2f^2-4V^4)>0$, i.e. $af>2V^2$.

Therefore collision will only occur if $af>2V$ and

$$\boldsymbol{m=\frac{2V^2}{(a^2f^2-4V^4)^{\frac{1}{2}}}}.$$

From above, **collision occurs at time**

$$t=\frac{2(1+m^2)^{\frac{1}{2}}}{f}V$$

$$=\frac{2V}{f}\left(1+\frac{4V^4}{(a^2f^2-4V^4)}\right)^{\frac{1}{2}}$$

$$=\frac{\mathbf{2Va}}{(\boldsymbol{a^2f^2-4V^4})^{\frac{1}{2}}}.$$

Example 2

(i) A particle, P, moves in a horizontal plane about a fixed point, O, with constant angular velocity, ω, and with zero acceleration along OP. Initially the distance OP is a and the radial velocity is $3a\,\omega$. Find the distance OP at time t.

(ii) A particle, P, of mass m lies on a smooth table and is attached by a string of length l, passing through a small hole O on the table, to a particle of mass M hanging vertically. When at a distance a $(<l)$, from the hole, P is projected along the table at right angles to the string, with speed u. Find an expression for the tension in the string when P is at a distance r from the hole, in terms of m, M, a, u and r.

Refer to figure 21(i).

In terms of the polar coordinates (r,θ) of P, we may use the standard results for the radial and transverse components of acceleration. With the usual notation,

radial component of acceleration $=\ddot{r}-r\dot{\theta}^2$

transverse component of acceleration $=2\dot{r}\dot{\theta}+r\ddot{\theta}$

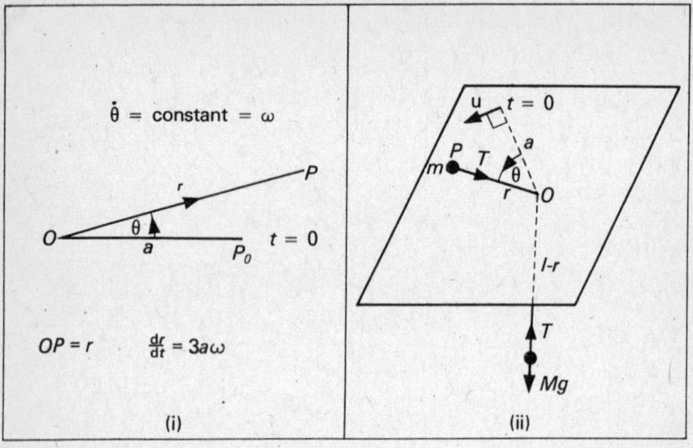

Figure 21

We are given that there is zero acceleration along OP, and that $\dot{\theta}=\omega=\text{constant}$,

i.e. $$\ddot{r}-r\dot{\theta}^2=0$$

$$\ddot{r}-r\omega^2=0$$

which is a **2nd order differential equation with constant coefficients**.

The auxiliary equation is

$$P^2-\omega^2=0$$

$$P=\pm\omega.$$

Thus, the **general solution** is

$$r=Ae^{P_1t}+Be^{P_2t'}, \qquad r=Ae^{\omega t}+Be^{-\omega t}$$

where A and B are constants.

When $t=0$, $r=a$ and $\dfrac{dr}{dt}=3a\omega$.

Differentiating and substituting,

$$\frac{dr}{dt}=A\omega e^{\omega t}-B\omega e^{-\omega t}$$

and $a=A+B$, $3a\omega=A\omega-B\omega$.

Combining these equations,

$$A=2a \text{ and } B=-a.$$

Thus the general solution is

$$\mathbf{r=a(2e^{\omega t}-e^{-\omega t})}$$

(ii) Refer to figure 21(ii).

We shall consider the equations of motion for the particle.

Tangentially, there is no external force, and therefore there is zero acceleration.

Thus $$r\ddot{\theta}+2\dot{r}\dot{\theta}=0. \tag{i}$$

Radially, applying '$f=ma$',

$$-T=m(\ddot{r}-r\dot{\theta}^2). \tag{ii}$$

For the mass M, applying '$f=ma$' vertically,

$$Mg-T=M\left(\frac{d^2}{dt^2}(l-r)\right)$$

$$\ddot{r}=\frac{T-Mg}{M}. \tag{iii}$$

Combining equations (ii) and (iii),

$$(M+m)T=mM(g+r\dot{\theta}^2). \tag{iv}$$

From equation (i),

$$\frac{\ddot{\theta}}{\dot{\theta}}=-2\frac{\dot{r}}{r}, \qquad \frac{1}{\dot{\theta}}\frac{d\dot{\theta}}{dt}=-\frac{2}{r}\frac{dr}{dt}.$$

Integrating,

$$\int\frac{1}{\dot{\theta}}d\dot{\theta}=-2\int\frac{1}{r}dr$$

$$\ln\dot{\theta}=-2\ln r+\ln C \text{ where } C \text{ is a constant.}$$

$$\dot{\theta}=\frac{C}{r^2}.$$

We know that when $r=a$,

$$\dot{\theta}=\frac{u}{a}, \text{ thus } C=au \text{ and } \dot{\theta}=\frac{au}{r^2}.$$

Substituting into equation (iv),

$$(M+m)T=mM\left(g+\frac{ra^2u^2}{r^4}\right)$$

$$T=\frac{mM}{M+m}\left(g+\frac{a^2u^2}{r^3}\right).$$

[We could also have derived the expression for $\dot{\theta}$ from the fact that, since there is no external torque acting on the system, then angular momentum about an axis perpendicular to the plane of the table, through the hole O, is conserved.

i.e. $$mau=mr^2\dot{\theta}$$

$$\dot{\theta}=\frac{au}{r^2} \qquad].$$

Example 3

A particle moves so that its acceleration at any time t is given by

$$\ddot{\mathbf{r}}=-a\omega^2(\cos\omega t\,\mathbf{i}+\sin\omega t\,\mathbf{j})$$

where a and ω are positive constants. Given that, at time

$$t=\frac{2\pi}{\omega},$$

the particle has velocity $a\omega\mathbf{j}$ and is at the point with position vector $2a\mathbf{i}$, find its position vector, $\mathbf{r}$, at any time t. Hence show that

$$\ddot{\mathbf{r}}=-\omega^2(\mathbf{r}-a\mathbf{i}).$$

Find $|\mathbf{r}-a\mathbf{i}|$ and the speed $|\dot{\mathbf{r}}|$ and show that $\mathbf{r}-a\mathbf{i}$ and $\dot{\mathbf{r}}$ are perpendicular. Briefly describe the motion of the particle.

At time, $t = \dfrac{4\pi}{\omega}$,

an impulse $ma\omega(2\mathbf{i}+\mathbf{j})$ is applied to the particle, which has mass m. Find the velocity of the particle immediately after the impulse and determine the change in its kinetic energy.

(A.E.B.)

The acceleration of the particle is given by

$$\ddot{\mathbf{r}} = -a\omega^2(\cos\omega t\,\mathbf{i} + \sin\omega t\,\mathbf{j})$$

where a, ω are positive constants.

Integrating with respect to time,

$$\dot{\mathbf{r}} = -a\omega^2 \int \cos\omega t\ \mathbf{i} + \sin\omega t\ \mathbf{j}\,\mathrm{d}t$$

$$= -a\omega^2\left(\frac{1}{\omega}\sin\omega t\,\mathbf{i} - \frac{1}{\omega}\cos\omega t\,\mathbf{j}\right) + \mathbf{c}$$

where $\mathbf{c}$ is a constant.

We are given that at $t = \dfrac{2\pi}{\omega}$, $\dot{\mathbf{r}} = a\omega\,\mathbf{j}$.

Therefore,

$$a\omega\mathbf{j} = -a\omega\mathbf{j} + \mathbf{c}, \quad \mathbf{c} = \mathbf{0},$$

and $$\dot{\mathbf{r}} = -a\omega(\sin\omega t\,\mathbf{i} - \cos\omega t\,\mathbf{j}).$$

Integrating with respect to time,

$$\mathbf{r} = -a\omega\left(-\frac{1}{\omega}\cos\omega t\,\mathbf{i} - \frac{1}{\omega}\sin\omega t\,\mathbf{j}\right) + \mathbf{d}$$

where $\mathbf{d}$ is constant.

We are given that at $t = \dfrac{2\pi}{\omega}$, $\mathbf{r} = 2a\mathbf{i}$.

Therefore,

$$2a\mathbf{i} = a\mathbf{i} + \mathbf{d}, \ \mathbf{d} = a\mathbf{i}$$

and $$\mathbf{r} = a(\cos\omega t\,\mathbf{i} + \sin\omega t\,\mathbf{j}) + a\mathbf{i}.$$

Hence $$\ddot{\mathbf{r}} = -\omega^2 \mathbf{r} - \omega^2 a\mathbf{i}$$
$$= -\omega^2(\mathbf{r} - a\mathbf{i})$$
$$|\mathbf{r} - a\mathbf{i}| = |a(\cos \omega t\,\mathbf{i} + \sin \omega t\,\mathbf{j})|$$
$$= a$$

$$|\dot{\mathbf{r}}| = |a\omega(\sin \omega t\,\mathbf{i} - \cos \omega t\,\mathbf{j})|$$
$$= a\omega.$$

In order to show that $(\mathbf{r} - a\mathbf{i})$ and $\dot{\mathbf{r}}$ are perpendicular, we shall inspect the scalar product of these two vectors.

$$(\mathbf{r} - a\mathbf{i}) . \dot{\mathbf{r}}$$
$$= (a \cos \omega t\,\mathbf{i} + a \sin \omega t\,\mathbf{j}) . (-a\omega \sin \omega t\,\mathbf{i} + a\omega \cos \omega t\,\mathbf{j})$$
$$= a^2\omega(-\cos \omega t \sin \omega t + \sin \omega t \cos \omega t) = 0.$$

Thus, **these two vectors are perpendicular**.

Refer to figure 22.

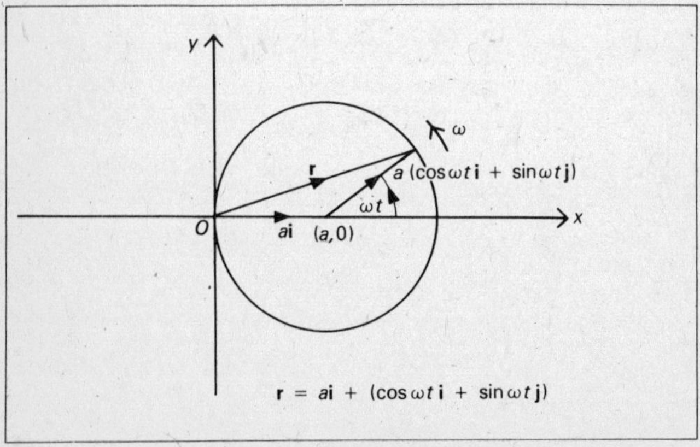

Figure 22

The motion of the particle is on a circular path, radius a, centre $(a, 0)$ with constant angular velocity ω.

At time $t = \dfrac{4\pi}{\omega}$,

an impulse $\mathbf{J} = ma\,\omega(2\mathbf{i} + \mathbf{j})$ is applied to the particle.

Refer to figure 23.

Applying 'impulse = change in momentum',

in the Ox direction,

$$2ma\omega = mV_x - 0$$
$$V_x = 2a\omega.$$

in the Oy direction,

$$ma\omega = mV_y - ma\omega$$
$$V_y = 2a\omega$$

Therefore **the velocity of the particle after the impulse**

$$= \mathbf{2a\omega i + 2a\omega j}.$$

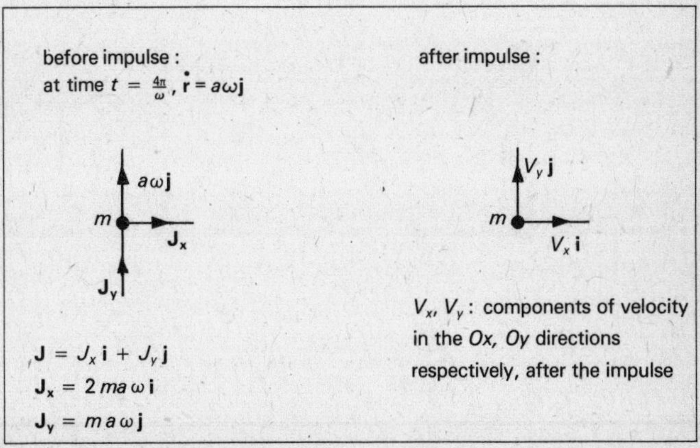

Figure 23

The K.E. before the impulse $=\frac{1}{2}ma^2\omega^2$.

The K.E. after the impulse $=\frac{1}{2}m4a^2\omega^2 2$.

Thus **the change in K.E.** $=\frac{7}{2}\boldsymbol{ma^2\omega^2}$.

Chapter 6
Projectiles

Example 1

A particle is projected from a point O on a cliff, at a speed of $29\,\text{ms}^{-1}$, and moves freely under gravity. The point O is at a vertical height of 60 metres above the sea and the particle strikes the sea at a point A, 6 seconds after the instant of its projection. Calculate:
(i) The sine of the angle of elevation at which the particle is projected
(ii) the horizontal distance between O and A.
At the instant when the particle is at the point B, where OB is horizontal, a second particle is projected from O and strikes the sea at A, simultaneously with the first particle. Calculate:
(iii) the time of flight of the second particle
(iv) the magnitude and the direction of the velocity of projection of the second particle.
(Take the acceleration due to gravity to be $10\,\text{ms}^{-2}$.)

(A.E.B.)

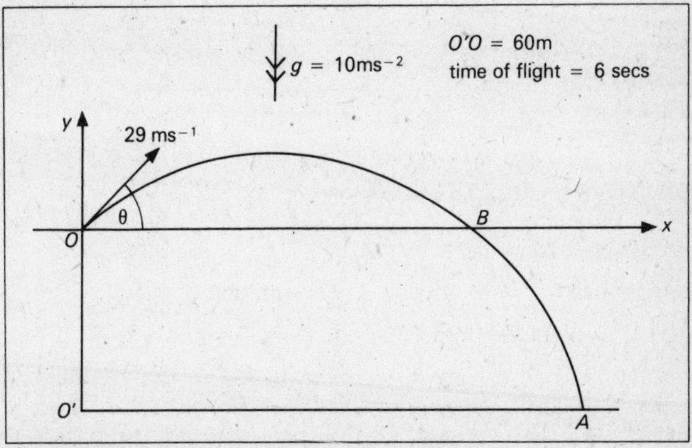

Figure 24

Refer to figure 24.

Taking the axes as shown,

$$\ddot{y} = -g = -10 \qquad \ddot{x} = 0$$

$$\dot{y} = 29 \sin\theta - 10t \qquad \dot{x} = 29\cos\theta$$

$$y = 29 \sin\theta\, t - 5t^2 \qquad x = 29\cos\theta\, t$$

(i) When $y = -60$, $t = 6$

Substituting, $-60 = 29\sin\theta\, 6 - 5.6.6$

$$\mathbf{\sin\theta = \frac{20}{29}} \text{ and } \mathbf{\cos\theta = \sqrt{\frac{29^2 - 20^2}{29}} = \frac{21}{29}}.$$

(ii) The horizontal distance between O and A is the value of x when t is 6 seconds, (i.e. the time of flight).

Thus $\boldsymbol{OA} = 29\cos\theta\, 6 = \mathbf{126\,m}$.

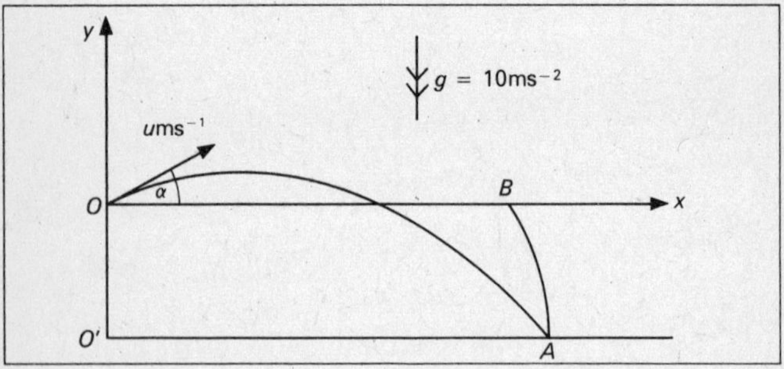

Figure 25

Referring to figure 25.

(iii) When the first particle is at B, $y = 0$.

Substituting, $0 = 29\sin\theta\, t - 5t^2 = t(t-4)$.

Hence, $t = 0$ (the initial position), or $t = 4$ (when it is at B).

Since the second particle leaves O at the instant the first particle passes through B, and reaches A at the same instant as the first particle, then **the time of flight for the second particle** is $(6-4)$ secs $=$ **2 s**.

(iv) We know that the time of flight of the second particle is 2 s and that its range is 126 m. Taking axes as shown, initial velocity u, and angle of projection α:

$$\ddot{y} = -g \qquad \ddot{x} = 0$$

$$\dot{y} = u \sin\alpha - 10t \qquad \dot{x} = u\cos\alpha$$

$$y = u\sin\alpha\, t - 5t^2 \qquad x = u\cos\alpha\, t$$

Hence, $126 = u\cos\alpha\, 2$, $\boldsymbol{u\cos\alpha = 63}$ and $-60 = u\sin\alpha\, 2 - 20$, $\boldsymbol{u\sin\alpha = -20}$.

Therefore $\tan\alpha = -\dfrac{20}{63}$, i.e. $\boldsymbol{\alpha = \arctan\dfrac{20}{63}}$

below the horizontal.

Since $\sin^2\alpha + \cos^2\alpha = 1$, then

$$20^2 + 63^2 = u^2$$

and $\boldsymbol{u = 66.1\ \mathrm{ms}^{-1}}$.

Example 2

A plane is inclined at α to the horizontal. A particle is projected up this plane with a velocity of projection V at θ to the horizontal. Show that it reaches a distance

$$\frac{2V^2\cos\theta\sin(\theta-\alpha)}{g\cos^2\alpha}$$

up the plane.

A particle is now projected down the same plane, with the same velocity of projection, V, at θ to the horizontal. The distance travelled down the plane is twice the distance travelled up the plane.

Prove $\tan\alpha = \frac{1}{3}\tan\theta$.

(S.U.J.B.)

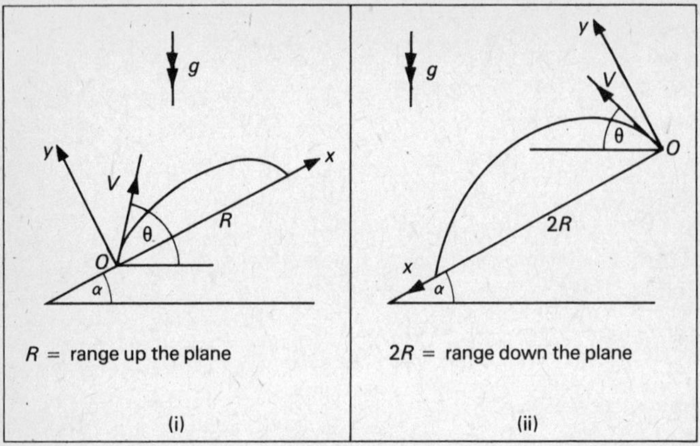

Figure 26

Refer to figure 26(i).

Take the axes as shown in the diagram, and applying the standard equations of motion for **constant** acceleration,

$$\ddot{y} = -g\cos\alpha \qquad \text{(i)}$$

$$\dot{y} = v\sin(\theta-\alpha) - g\cos\alpha\, t \qquad \text{(ii)}$$

$$y = v\sin(\theta-\alpha)\,t - \frac{g}{2}\cos\alpha\, t^2 \qquad \text{(iii)}$$

$$\ddot{x} = -g\sin\alpha \qquad \text{(iv)}$$

$$\dot{x} = v\cos(\theta-\alpha) - g\sin\alpha\, t \qquad \text{(v)}$$

$$x = v\cos(\theta-\alpha)\,t - \frac{g}{2}\sin\alpha\, t^2. \qquad \text{(vi)}$$

The particle will have travelled its **range** up the plane, touching the plane again **when $y=0$** (after the initial $y=0$, $t=0$).

Substituting in (iii),

$$0 = V\sin(\theta-\alpha)\,t - \tfrac{1}{2}g\cos\alpha\, t^2.$$

Thus, either $t=0$ (initial position),

or $\boldsymbol{t}=\dfrac{\mathbf{2}\boldsymbol{v}\,\mathbf{sin}\,(\boldsymbol{\theta}-\boldsymbol{\alpha})}{\boldsymbol{g}\,\mathbf{cos}\,\boldsymbol{\alpha}}$ which is the **time of flight**.

In order to find the range, R, we substitute this value of t into equation (vi).

$$\boldsymbol{R}=v\cos(\theta-\alpha)\frac{2v\sin(\theta-\alpha)}{g\cos\alpha}-\frac{1}{2}g\sin\alpha\left(\frac{2v\sin(\theta-\alpha)}{g\cos\alpha}\right)^2$$

$$=\frac{2v^2}{g\cos^2\alpha}\sin(\theta-\alpha)\left[\cos(\theta-\alpha)\cos\alpha-\sin(\theta-\alpha)\sin\alpha\right]$$

$$=\frac{\mathbf{2}\boldsymbol{v}^2\,\mathbf{cos}\,\boldsymbol{\theta}\,\mathbf{sin}\,(\boldsymbol{\theta}-\boldsymbol{\alpha})}{\boldsymbol{g}\,\mathbf{cos}^2\,\boldsymbol{\alpha}}.$$

The particle is now projected **down** the plane at the same speed, V at the same angle, θ, to the horizontal.

Refer to figure 26(ii).

Taking the axes as shown,

$$\ddot{y}=-g\cos\alpha \qquad \text{(vii)}$$

$$\dot{y}=v\sin(\theta+\alpha)-g\cos\alpha\,t \qquad \text{(viii)}$$

$$y=v\sin(\theta+\alpha)t-\tfrac{1}{2}g\cos\alpha\,t^2 \qquad \text{(ix)}$$

$$\ddot{x}=+g\sin\alpha \qquad \text{(x)}$$

$$\dot{x}=v\cos(\theta+\alpha)+g\sin\alpha\,t \qquad \text{(xi)}$$

$$x=v\cos(\theta+\alpha)t+\tfrac{1}{2}g\sin\alpha\,t^2 \qquad \text{(xii)}$$

Using a similar method to the first part of the question, the time of flight t, is given by

$$t=\frac{2v\sin(\theta+\alpha)}{g\cos\alpha}$$

and the range, $2R$, is given by

$$\boldsymbol{2R}=\frac{v\cos(\theta+\alpha)\,2v\sin(\theta+\alpha)}{g\cos\alpha}+\frac{1}{2}g\sin\alpha\left(\frac{2v\sin(\theta+\alpha)}{g\cos\alpha}\right)^2.$$

$$= \frac{2v^2 \sin(\theta+\alpha)\cos\theta}{g\cos^2\alpha}.$$

Comparing this result with the expression obtained for R from the first part, we see that

$$2\left(\frac{2v^2\cos\theta\sin(\theta-\alpha)}{g\cos^2\alpha}\right)=\frac{2v^2\sin(\theta+\alpha)\cos\theta}{g\cos^2\alpha}.$$

Thus,

$$2\sin(\theta-\alpha)=\sin(\theta+\alpha)$$

$$\sin\theta\cos\theta=3\sin\alpha\cos\theta$$

$$\tan\alpha=\frac{1}{3}\tan\theta.$$

Example 3

When serving in a ball game, a player projects the ball in a vertical plane perpendicular to the net with speed V from a height $2h$ above the ground. He stands at a distance $13h$ from the net, which is of height h. The serve is 'in' if the ball goes over the net and hits the ground within a distance $7h$ of the net. It is assumed that the ball may be treated as a particle and air-resistance ignored.

(i) The ball is projected horizontally. Show that the serve will be 'in' if

$$13\sqrt{(\tfrac{1}{2}gh)}<V<10\sqrt{(gh)}$$

(ii) The ball is projected at an angle α to the horizontal and $V=\sqrt{(13gh)}$. Show that the ball will clear the net if

$$1-\sqrt{\frac{2}{13}}<\tan\alpha<1+\sqrt{\frac{2}{13}}$$

and verify that if this condition is satisfied, the serve will be 'in'.

(W.J.E.C.)

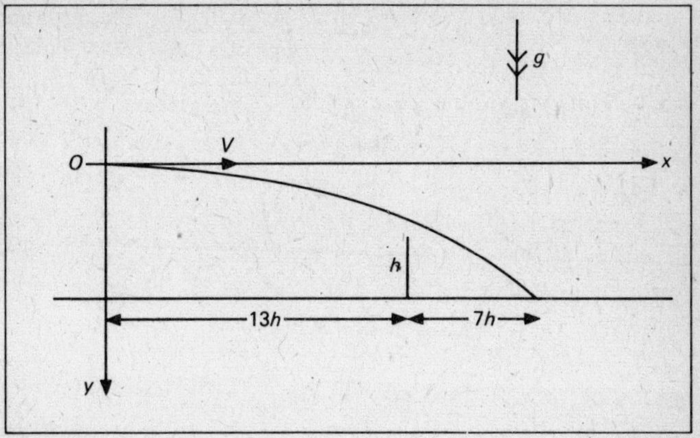

Figure 27

(i) Refer to figure 27.

The particle is projected horizontally.

Taking the axes as shown in the diagram, the equations of motion are derived from the standard equations for **constant acceleration** in the y-direction and **constant velocity** in the x-direction.

Thus,

$$\ddot{y}=g \qquad \ddot{x}=0$$

$$\dot{y}=gt \qquad \dot{x}=V$$

$$y=\tfrac{1}{2}gt \qquad x=Vt$$

For the serve to be 'in' it must (a) clear the net and (b) fall within a distance of $7h$ of the net.

Condition (a) requires that when $x=13h$, $y<h$.

Thus, when $(x=)\ 13h=Vt$, $\quad t=\dfrac{13h}{V}$

then $\qquad (y=)\ \tfrac{1}{2}gt^2<h$

i.e. $\qquad V^2>\dfrac{169}{2}gh.$

V is not negative,

thus, $\qquad \mathbf{V>13\sqrt{\dfrac{gh}{2}}}.$

Condition (b) requires that when the ball hits the ground, $x<20h$.

The ball hits the ground when $y=2h$,

i.e. when $$2h=\tfrac{1}{2}gt^2$$

$$t=2\sqrt{\frac{h}{g}}.$$

Thus, when $t=2\sqrt{\dfrac{h}{g}}$, $(x=)\,Vt<20h$

i.e. $$2V\sqrt{\frac{h}{g}}<20h$$

$$\boldsymbol{V<10\sqrt{gh}}.$$

Therefore, the conditions for the serve to be 'in' are

$$\boldsymbol{13\sqrt{\frac{gh}{2}}<V<10\sqrt{gh}}.$$

(ii) Refer to figure 28(i).

Taking the axes as shown in the diagram, the equations of motion are:

$\ddot{y}=g$	(i)	$\ddot{x}=0$	(iv)
$\dot{y}=-V\sin\alpha+gt$	(ii)	$\dot{x}=V\cos\alpha$	(v)
$y=-V\sin\alpha\,t+\frac{1}{2}gt^2$	(iii)	$x=V\cos\alpha\,t$	(vi)

The ball will clear the net if when $x=13h$ then $y<h$.

When $$(x=)\,13h=V\cos\alpha\,t$$

$$t=\sqrt{\frac{13h}{g}}\sec\alpha.$$

Thus to clear the net,

$$(y=)-\sqrt{13gh}\sin\alpha\sqrt{\frac{13h}{g}}\sec\alpha+\tfrac{1}{2}g\frac{13h}{g}\sec^2\alpha<h.$$

Rearranging and simplifying,

$$f(\tan\alpha)=13\tan^2\alpha-26\tan\alpha+11<0.$$

When $f(\tan\alpha)=0$,

then $$\tan\alpha=\frac{26\pm\sqrt{26^2-4.11.13}}{26}$$

$$=1\pm\sqrt{\frac{2}{13}}.$$

Refer to figure 28(ii).

Thus, for $f(\tan\alpha)<0$, i.e. **for the ball to clear the net**, then

$$1-\sqrt{\frac{2}{13}}<\tan\alpha<1+\sqrt{\frac{2}{13}}.$$

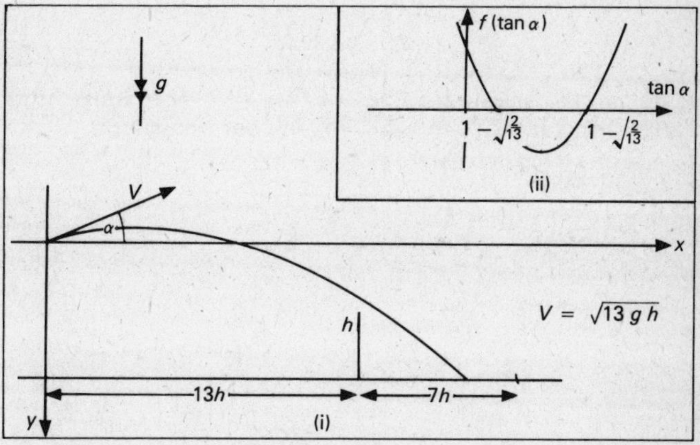

Figure 28

It is the case that the **maximum range for any velocity of projection occurs when the angle of projection is 45°.**

Within the above range for $\tan\alpha$ such that the ball clears the net, $\tan\alpha=1$ is included.

Thus the **maximum value** that x can take if it clears the net **occurs when $\tan\alpha = 1$**. In order to verify that the ball will be 'in' we examine the value of x when $\tan\alpha = 1$ and $y = 2h$.

Combining equations (iii) and (vi) to give **the equation of path**,

$$y = -x\tan\alpha + \tfrac{1}{2}g\frac{x^2}{V^2}\sec^2\alpha$$

and substituting,

$$y = 2h, \tan\alpha = 1, (\sec^2\alpha = 2), V = \sqrt{13gh}$$

gives

$$2h = -x + \frac{gx^2}{13gh}$$

$$x^2 - 13hx - 26h^2 = 0$$

$$x = \left(\frac{13 \pm \sqrt{13^2 + 104}}{2}\right)h$$

$$= (6.5 \pm 8.26)h.$$

Rejecting the negative value, we see that **the maximum range for x is 14.76 h, well 'in' the service range.**

Chapter 7
Work, Power and Energy

Example 1

A car of mass 1.2 tonnes is travelling up a slope of 1 in 150 at a constant speed of $10\,\text{ms}^{-1}$. If the frictional and air resistances are 100 newtons, calculate the power exerted by the engine. The car descends the same slope working at a rate of 2 kW. What will be its acceleration when its speed is $20\,\text{ms}^{-1}$ if the resistances are the same? If the engine is shut off when the speed of the car is $25\,\text{ms}^{-1}$ as it descends the slope, how long will it be before the car comes to rest? ($g=10\,\text{ms}^{-2}$).

(S.U.J.B.)

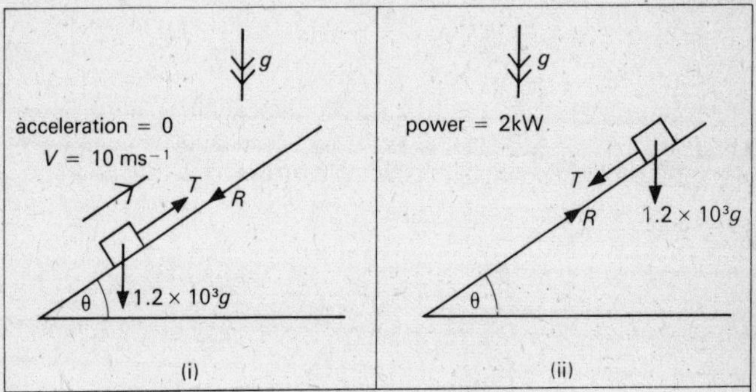

$\sin\theta = \frac{1}{150}$ resisting forces $= R = 100$ N

$g = 10\,\text{ms}^{-2}$ tractive force $= T = \dfrac{\text{power}}{\text{velocity}}$

Figure 29

Refer to figure 29(i).

Let the power of the engine be P, and the tractive force be T. Since 'power = tractive force × velocity' and the velocity is constant at $10\,\text{ms}^{-1}$,

$$T=\frac{P}{10}.$$

Applying Newton's second law, '$f=ma$' parallel to the slope:

$$\frac{P}{10}-100-1.2\times 10^3\times 10\times\frac{1}{150}=0$$

(the velocity is constant, therefore the acceleration is zero).

Hence, $\boldsymbol{P=1.8\times 10^3}$**W.**

Refer to figure 29(ii).

Applying Newton's second law parallel to the slope, when the velocity is 20 ms^{-1}, and the acceleration is a:

$$\frac{2\times 10^3}{20}-100+1.2\times 10^3\times 10\times\frac{1}{150}=1.2\times 10^3 a.$$

Hence, $\boldsymbol{a=\frac{1}{15}}$ $\textbf{ms}^{-2}$.

When the speed of the car is 25 ms^{-1} the engine is shut off and will move under the constant resisting forces and its component of weight down the slope, producing **constant** acceleration f.

Applying '$f=ma$',

$$1.2\times 10^3\times 10\times\frac{1}{150}-100=1.2\times 10^3 f$$

$$\boldsymbol{f=-\frac{1}{60}}\ \textbf{ms}^{-2}.$$

Applying the equation of motion '$v=u+at$' for constant acceleration:

$$0=25-\frac{1}{60}t,\quad t=25\times 60\ \text{secs}.$$

Therefore **the car takes 25 minutes to come to rest.**

Example 2

A water pump raises 50 kg of water per second through a height of 20 m. The water emerges as a jet with speed $50\,ms^{-1}$. Find the actual power of the pump if it is 80% efficient. The jet is directed at 30° above the horizontal, find the further height attained by the water.

The jet impinges at its highest point on a vertical wall. Find the least force that is exerted on the wall.

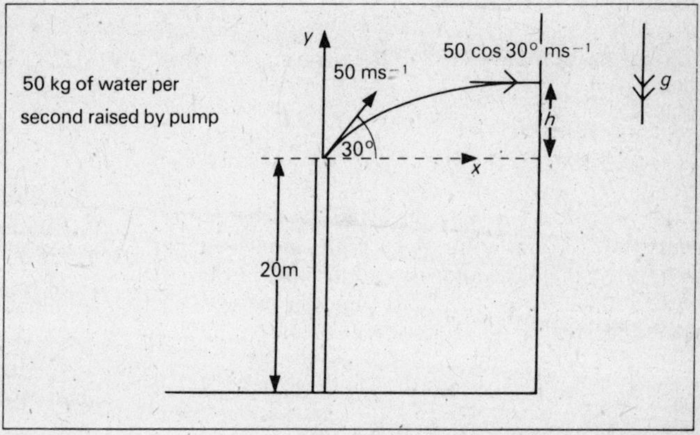

Figure 30

Refer to figure 30.

The work done against gravity per second in raising 50 kg of water through $20\,m = 50 \times g \times 20$

$$= 10\,000\,W = \mathbf{10\,kW}$$

The work done per second in giving 50 kg of water a velocity of $50\,ms^{-1} = \frac{1}{2} \times 50 \times 50^2$

$$= 62{,}500\,W = \mathbf{62.5\,kW}.$$

Thus **the effective power of the pump** $= 10 + 62.5$

$$= \mathbf{72.5\,kW}.$$

(i.e. the effective power of the pump = the sum of the kinetic energy per second and the potential energy per second given to the water.)

The pump is **80% efficient**, therefore **the actual power of the pump**

$$= 72.5 \times \frac{100}{80} = \mathbf{90.6\,kW}.$$

The jet is directed at 30° to the horizontal and impinges on the wall when it is at its highest point, i.e. when it is instantaneously travelling **horizontally**.

Taking the axes as shown in the diagram,

$$\ddot{y} = -g \qquad \ddot{x} = 0$$
$$\dot{y} = 50 \sin 30^\circ - gt \qquad \dot{x} = 50 \cos 30^\circ$$
$$y = 50 \sin 30^\circ\, t - \tfrac{1}{2} gt \qquad x = 50 \cos 30^\circ\, t$$

The force exerted on the wall by the water is equal in magnitude to the amount of momentum lost per second. The water at its highest point is travelling with speed $50 \cos 30^\circ\ \mathrm{ms^{-1}}$ horizontally. Thus the least amount of momentum lost per second (assuming no rebound) = Force exerted on the wall,

$$= 50 \times 50 \cos 30^\circ = 1250\sqrt{3}\ \mathrm{N}.$$

The water reaches its **highest point, *h***, when $\dot{y} = 0$, i.e.

when $0 = 50 \sin 30^\circ - 10t$, $\quad \boldsymbol{t = \frac{5}{2}\,\mathrm{s}}$.

At this time, $\qquad y = \boldsymbol{h} = 50 \sin 30^\circ \frac{5}{2} - \frac{10}{2}\frac{25}{4}$

$$= \mathbf{31.25\,m}.$$

Chapter 8
Momentum, Impulse and Impact

Example 1

Three balls, A, B and C, with masses m_1, m_2 and m_3, respectively, are at rest in a straight line on a horizontal plane. A is projected with velocity u to hit B, and A immediately stops. B carries on, hitting C, and B immediately stops. C carries on, hitting a vertical wall at right angles, rebounds, striking B again, which goes on to hit A.

If e is the coefficient of restitution between any two of the balls, and between C and the wall,

(i) prove that $m_2 = \frac{1}{e} m_1$ and $m_3 = \frac{1}{e^2} m_1$

(ii) find the speeds of the three balls (in terms of u and e only) when all are in motion.

(S.U.J.B.)

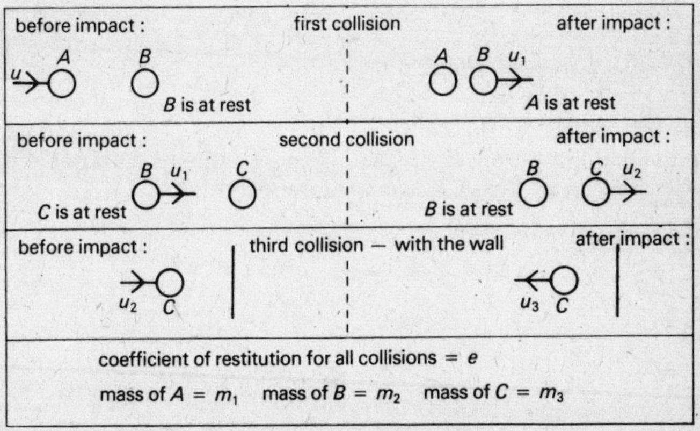

Figure 31

Refer to figure 31.

We shall apply the **Principle of Conservation of Linear Momentum** and **Newton's Law of Restitution** along the line of impact, for each of the collisions.

(i) First Collision.

Conservation of Momentum:

$$m_1 u = m_2 u_1.$$

Newton's Law:

$$u_1 = eu$$

Combining these equations,

$$\boldsymbol{m_2 = \frac{1}{e} m_1}.$$

Second Collision.

Conservation of Momentum:

$$m_2 u_1 = m_3 u_2$$

Newton's Law:

$$u_2 = eu_1$$

Combining these equations and substituting for m_2,

$$m_2 = em_3, \quad \boldsymbol{m_3 = \frac{1}{e^2} m_1}.$$

(ii) Third Collision.

In this collision, the wall is not free to move and an external impulse will be applied to the system. Hence we may not use the Principle of Conservation of Linear Momentum.

Newton's Law:

$$u_3 = -eu_2, \text{ hence } u_3 = -e^3 u.$$

Refer to figure 32.

Fourth Collision.

Conservation of Momentum:

$$m_3 u_3 = m_3 u_4 + m_2 u_5.$$

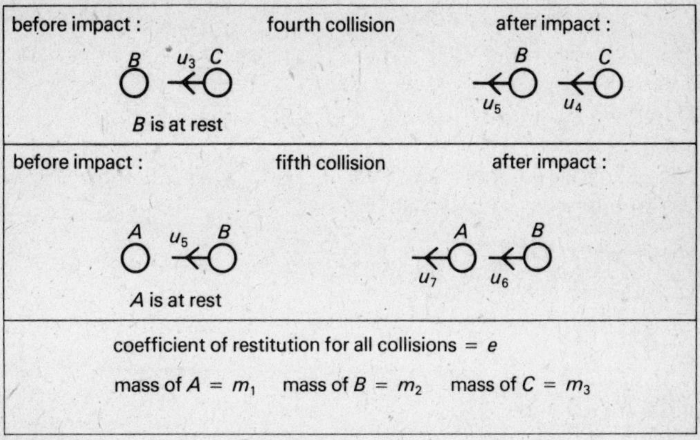

Figure 32

Substituting for m_2, m_3, u_3,

$$-ue^3 = u_4 + eu_5$$

Newton's Law:

$$\frac{u_5 - u_4}{0 - u_3} = -e, \quad u_5 - u_4 = eu_3.$$

Hence, $$u_5 - u_4 = -e^4u.$$

Combining these equations,

$$u_5(1+e) = -e^3u(1+e). \text{ Hence, } u_5 = -e^3u$$

and $$\boldsymbol{u_4 = e^3(e-1)u}.$$

Fifth Collision.

Conservation of Momentum:

$$m_2u_5 = m_2u_6 + m_1u_7.$$

Substituting for m_1, m_2, u_5,

$$-ue^3 = u_6 + eu_7.$$

Newton's Law:

$$\frac{-u_7 - u_6}{0 - u_5} = -e, \text{ i.e. } u_7 - u_6 = eu_5 = -e^4u.$$

Combining these equations,

$$u_7(1+e) = -e^3u(1+e).$$

Thus, $$\boldsymbol{u_7 = -e^3u} \text{ and } \boldsymbol{u_6 = e^3(e-1)u}.$$

The velocities of the three balls A, B, C when all three are in motion are u_7, u_6, u_4 respectively i.e.

$$\boldsymbol{-e^3u, -e^3(1-e)u, -e^3(1-e)u}.$$

Example 2

Two equal smooth spheres are moving in opposite directions, one with twice the speed of the other. They collide obliquely, their directions of motion making an angle of $\pi/3$ with the line of the centres at impact. The coefficient of restitution for the collision is $\frac{2}{3}$.

Find the magnitude and direction of the velocities of each of the spheres after the impact and the percentage loss of the kinetic energy due to the collision.

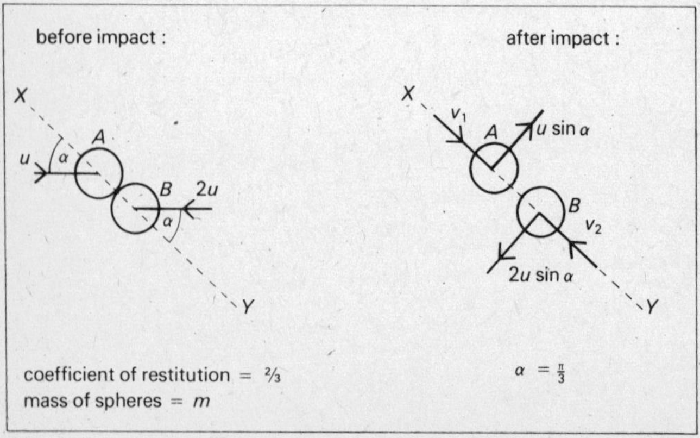

Figure 33

Refer to figure 33.

Along the line of centres, no external impulse is acting. Thus we can apply the **Principle of the Conservation of Linear Momentum**:

$$mu\cos\frac{\pi}{3}-2mu\cos\frac{\pi}{3}=mV_1-mV_2$$

i.e. $$V_2-V_1=\frac{u}{2}.$$

Newton's Law of Restitution along the line of impact:

$$\frac{V_1+V_2}{u\cos\pi/3+2u\cos\pi/3}=\frac{-2}{3},\text{ i.e. } V_1+V_2=-u.$$

Combining these equations,

$$V_2=-\frac{u}{4},\ V_1=-\frac{3}{4}.$$

Perpendicular to the line of impact, no impulsive force acts and the momenta of the spheres will remain unaltered in that direction, as shown in figure 33.

The magnitude of the velocity of *A* after impact

$$=\sqrt{\frac{9}{16}u^2+\frac{3}{4}u^2}=u\sqrt{\frac{21}{4}}.$$

The magnitude of the velocity of *B* after impact

$$=\sqrt{\frac{u^2}{16}+3u^2}=7\frac{\mathbf{u}}{\mathbf{4}}.$$

The direction of the velocity of *A* after impact

$$=\arctan\frac{u\sqrt{3}/2}{-V_1}\text{ with }YX=\mathbf{arctan}\frac{\mathbf{2}}{\sqrt{\mathbf{3}}}\text{ with }YX.$$

The direction of the velocity of *B* after impact

$$=\arctan\frac{u\sqrt{3}}{-V_2}\text{ with }XY=\mathbf{arctan\,4\sqrt{3}}\text{ with }XY.$$

Example 3

A uniform smooth sphere, A of mass $3m$ and centre C is at rest on a smooth horizontal table. It is attached to a fixed point O by a horizontal string of length l. Initially the string is taut. A second uniform smooth sphere, B, of mass $2m$ and identical radius to that of A, is moving with speed u in the direction **CO** towards A. On collision, the line of centres is at 60° to the direction of B's motion. The coefficient of restitution is $\frac{1}{4}$. Show that the speed of A after the collision is $u/4$ and find B's speed after the collision. Show also that the angular speed of the string after it becomes taut is

$$\frac{u\sqrt{3}}{8l}.$$

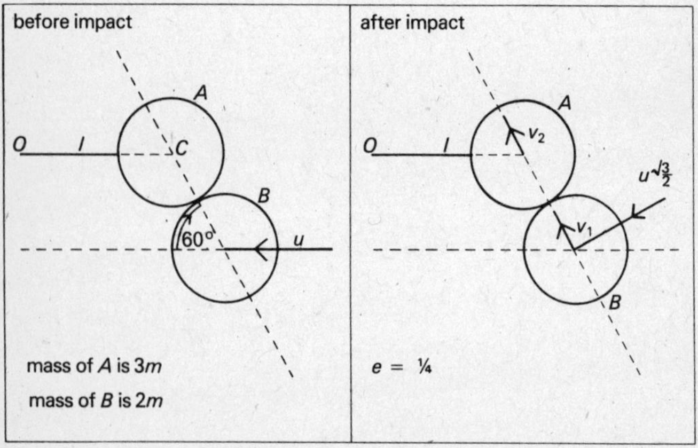

Figure 34

Refer to figure 34.

Motion perpendicular to the line of centres is unchanged. Thus, after impact, B still has velocity $u\sqrt{3}/2$ in this direction.

Applying **Newton's Law of Restitution** along the line of centres,

$$\frac{v_1 - v_2}{u/2 - 0} = -\frac{1}{4}, \text{ thus, } v_1 - v_2 = -\frac{u}{8}.$$

Momentum is conserved along the line of centres, therefore

$$2m\frac{u}{2} = 3mv_2 + 2mv_1.$$

Combining these equations, $\boldsymbol{v}_1 = \frac{\boldsymbol{u}}{\mathbf{8}}$ and $\boldsymbol{v}_2 = \frac{\boldsymbol{u}}{\mathbf{4}}$.

Hence, **after the collision the velocity of *A* is $\frac{\boldsymbol{u}}{\mathbf{4}}$ in the direction shown, and the speed of *B* is**

$$\sqrt{\frac{u^2}{64} + u^2\frac{3}{4}} = \frac{\mathbf{7}\boldsymbol{u}}{\mathbf{8}}.$$

A continues with this velocity until the string becomes taut again.

Refer to figure 35.

Perpendicular to the string there is no impulse and the velocity in that direction will be unchanged by the impulsive tension as the string becomes taut.

Thus, $$v = \frac{u}{4}\cos 30° = \frac{u\sqrt{3}}{8}$$

and **the angular speed of the string after it becomes taut is**

$$\frac{\boldsymbol{u}\sqrt{\mathbf{3}}}{\mathbf{8}\boldsymbol{l}}.$$

[Alternatively, we could have used the fact that since no external torque is applied to the system then angular momentum about *O* is conserved, thus

$$l \sin 60° \frac{u}{4} = lv$$

and the same result follows.]

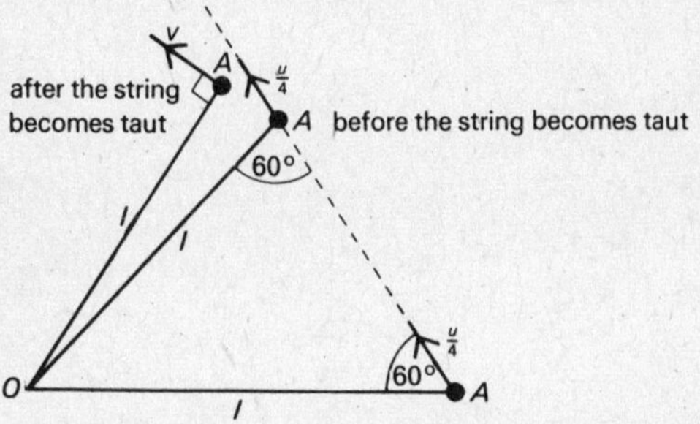

Figure 35

Example 4

A gun mass M is free to move in a straight line on a smooth horizontal track and is fired in the direction of the track. It fires a shell of mass kM along its barrel which is at an elevation α. Show that the initial direction of motion of the shell is inclined at an angle $\tan^{-1}\{(1+k)\tan\alpha\}$ to the horizontal.

If the whole of the energy generated by the explosion, $\frac{1}{2}Mn^2$, is transferred to the gun and shell, show that the initial speed of the gun is

$$n\left(1+\frac{1}{k}[1+(1+k)^2\tan^2\alpha]\right)^{-1/2}.$$

(W.J.E.C.)

Refer to figure 36.

Before the gun is fired, both the gun and shell are at rest; therefore the initial total momentum is zero. The gun is free to move on the smooth horizontal track. Hence **in this direction** there is no external force acting and **linear momentum is conserved**.

Therefore, applying the Principle of Conservation of Linear Momentum in the direction of recoil,

$$0 = MV + kM(V - v\cos\alpha),$$

and

$$V = \frac{kv\cos\alpha}{1+k}, \quad v = \frac{V(1+k)}{k\cos\alpha}$$

When the shell leaves the barrel, it has velocity, V_E, relative to earth, which is the resultant of its components v and V as shown in figure 36.

Thus V_E has horizontal component $(v\cos\alpha - V)$ and vertical component $v\sin\alpha$ and is inclined at angle

$$\arctan\frac{v\sin\alpha}{v\cos\alpha - V} = \arctan\left(\frac{v\sin\alpha}{v\cos\alpha - \frac{kv\cos\alpha}{1+k}}\right).$$

$$= \arctan((1+k)\tan\alpha) \text{ to the horizontal.}$$

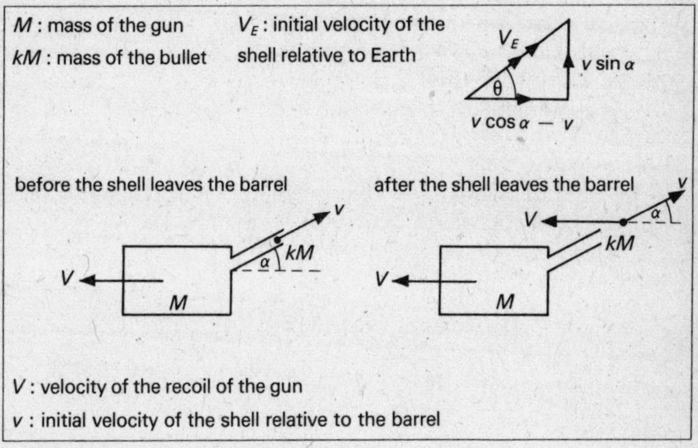

Figure 36

If the whole of the energy, $\frac{1}{2}Mn^2$, of the explosion is transferred to the shell and the gun then,

$$\tfrac{1}{2}Mn^2 = \tfrac{1}{2}MV^2 + \tfrac{1}{2}kM[v^2\sin^2\alpha + (v\cos\alpha - V)^2].$$

Thus,

$$n^2 = V^2 + k\left(\frac{V^2(1+k)^2\tan^2\alpha}{k^2}\right) + k\left(\frac{V}{k}(1+k) - V\right)^2$$

$$= V^2 + \frac{V^2}{k}(1+k)^2\tan^2\alpha + \frac{V^2}{k}$$

$$= V^2\left(1 + \frac{1}{k}(1 + (1+k)^2\tan^2\alpha)\right)$$

and $$\boldsymbol{V = n\left(1 + \frac{1}{k}(1 + (1+k)^2\tan^2\alpha)\right)^{-1/2}}$$

Example 5

Two particles A and B of masses $11m$ and $5m$ respectively are at rest on a smooth horizontal table. A light inextensible string AB joins them. The string is just taut. A horizontal impulse, J, is given to particle B so that B moves with initial speed V in a direction making angle $\arctan \frac{4}{3}$ with AB produced.

Calculate the magnitude and direction of J and show that the total kinetic energy generated by the impulse is

$$\frac{112}{25}mV^2.$$

Refer to figure 37.

When the impulse J is applied to particle B an impulsive tension, T is generated in the string AB.

We shall apply

'impulse=change in linear momentum'.

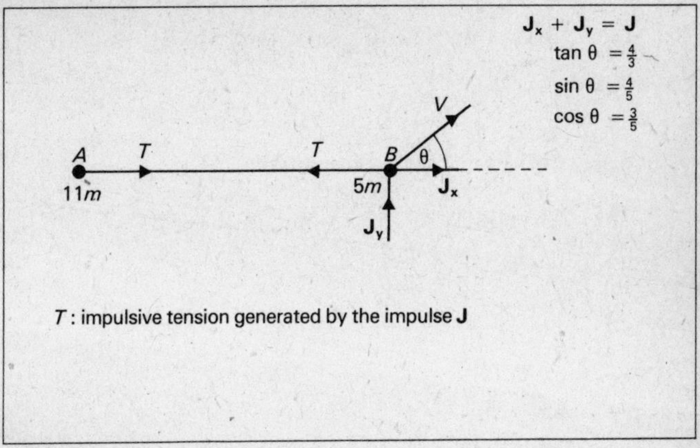

Figure 37

Parallel to the string,

particle B, $$J_x - T = 5mV\cos\theta$$
$$J_x = T + 3mV$$

particle A, $$T = 11mV\cos\theta$$

(A must move off in the direction of the impulsive tension, T, with the same speed as B has in that direction, otherwise the string would break.)

Perpendicular to the string,

$$J_y = 5mV\sin\theta$$
$$= 4mV$$

Combining these equations,

$$J_x = \frac{48}{5}mV, \quad J_y = 4mV$$

and $$J = \left(16m^2V^2 + \frac{48^2}{25}m^2V^2\right)^{\frac{1}{2}}.$$

Therefore, J has magnitude $\frac{52}{5}mV$ and direction $\arctan\frac{5}{12}$ to AB produced.

The K.E. produced by the impulse

$$=\text{initial K.E. of } A+\text{initial K.E. of } B$$

$$=\frac{1}{2}11m\frac{9}{25}V^2+\frac{1}{2}5mV^2$$

$$=\frac{112}{25}mV^2.$$

Chapter 9
Circular Motion of Particles

Example 1

A particle is moving in a circle, centre O and radius r, at a constant speed u. Prove that the acceleration of the particle is u^2/r directed towards O.

A car is moving in a horizontal circle of radius 100 metres on a track which is banked at an angle θ to the horizontal, where $\tan\theta = 9/40$. Calculate:
(i) the speed of the car when there is no tendency for it to slip sideways
(ii) the greatest steady speed at which the car could move in this circle without slipping sideways, given that the coefficient of friction between the car and the ground is 0.1.
(Take the acceleration due to gravity to be $10\,\text{ms}^{-2}$.)

(A.E.B.)

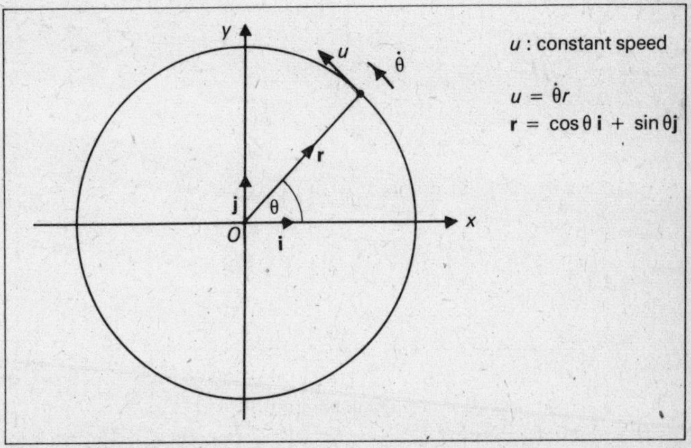

Figure 38

Refer to figure 38.

The position vector (relative to O), $\mathbf{r}$, of the particle moving in a circle, centre O, radius r, with constant speed u, is given by,

$$\mathbf{r}=r\cos\theta\,\mathbf{i}+r\sin\theta\,\mathbf{j}.$$

Differentiating w.r.t. time,

$$\dot{\mathbf{r}}=-r\sin\theta\,\dot{\theta}\,\mathbf{i}+r\cos\theta\,\dot{\theta}\,\mathbf{j}$$

$\dot{\theta}$ is constant.

Differentiating again,

$$\ddot{\mathbf{r}}=-r\dot{\theta}^2\cos\theta\,\mathbf{i}-r\dot{\theta}^2\sin\theta\,\mathbf{j}.$$

Thus, $$\ddot{\mathbf{r}}=-\dot{\theta}^2\,\mathbf{r} \text{ and } |\ddot{\mathbf{r}}|=\dot{\theta}^2\,|\mathbf{r}|=\frac{u^2}{r}.$$

Therefore, the acceleration of the particle, $\ddot{\mathbf{r}}$, **has magnitude**

$$\boldsymbol{\frac{u^2}{r}},$$

and is directed towards the centre of the circular path, *O*.

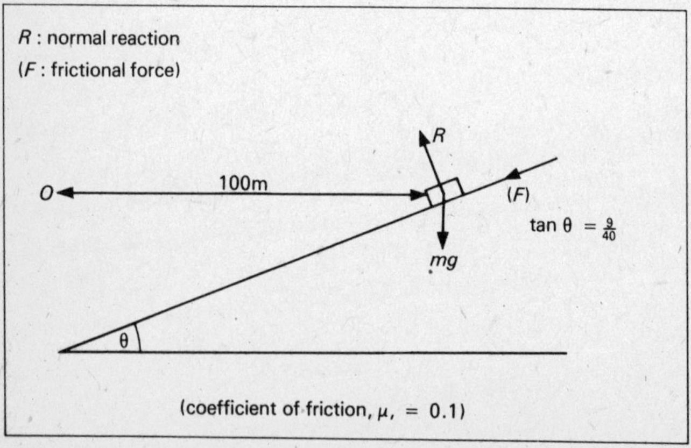

Figure 39

Refer to figure 39.

(i) When there is no tendency for the car to slip sideways, then no frictional force has been called into play to oppose potential slipping. Thus, applying Newton's second law, '$f=ma$', **radially**, (horizontally),

$$R\sin\theta=\frac{mV^2}{100}.$$

where V is the speed of the car for which there is no tendency to side-slip.

Vertically, there is no component of acceleration, i.e.

$$R\cos\theta=mg.$$

Combining these equations

$$V^2=100g\tan\theta,\quad V=\mathbf{15\,ms^{-1}}.$$

(ii) The greatest steady speed at which the car could move in this circle without slipping sideways, V_1, occurs when the **frictional force** opposing potential slipping is at its **limiting value**. Applying '$f=ma$', **radially**,

$$R\sin\theta+F\cos\theta=m\frac{V_1^2}{100} \qquad \text{(i)}$$

and **vertically**,

$$R\cos\theta=mg+F\sin\theta. \qquad \text{(ii)}$$

Since friction is limiting, $F=\mu R=\frac{R}{10}$. (iii)

Combining these equations:

Subst. (iii) in (i), $10F\sin\theta+F\cos\theta=\frac{mV_1^2}{100}$.

Subst. (iii) in (ii), $10F\cos\theta-F\sin\theta=10m$.

Dividing, $$\frac{10\sin\theta+\cos\theta}{10\cos\theta-\sin\theta}=\frac{V_1^2}{1000}.$$

Therefore, $$V_1^2=\frac{13}{391}\times10^4,\quad V_1=\mathbf{18.2\,ms^{-1}}.$$

Example 2

A particle P is attached to one end of a light inextensible string of length a whose other end is attached to a fixed point O. The particle is hanging with the string vertical when it is projected horizontally with speed $\sqrt{(2gb)}$.

Show that if

$$a < b < \frac{5}{2}a$$

the string becomes slack, and in this case show that the greatest height reached by P above its lowest point is

$$\frac{(4a-b)(a+2b)^2}{27a^2}.$$

(W.J.E.C.)

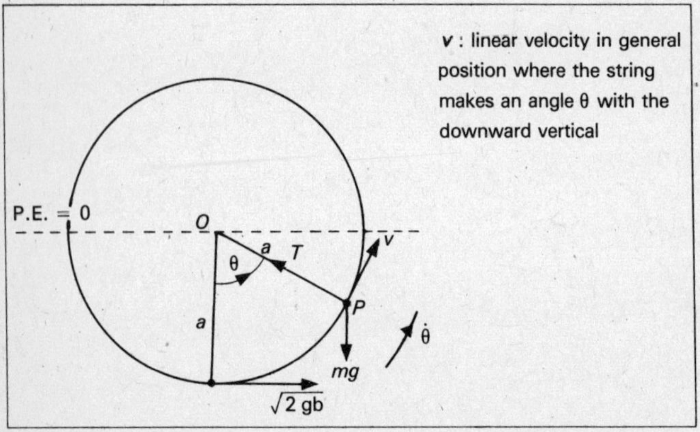

Figure 40

Refer to figure 40.

This is a conservative system of forces. Thus, by the **Principle of Conservation of Energy** (equating total mechanical energy in the initial position and the general position shown).

$$-mga+\tfrac{1}{2}m\,2gb=-mga\cos\theta+\tfrac{1}{2}mv^2$$

$$\boldsymbol{v^2=2g(a\cos\theta-a+b)}. \qquad \text{(i)}$$

Analysing the motion radially,

$$T-mg\cos\theta=m\frac{v^2}{a}. \qquad \text{(ii)}$$

Combining equations (i) and (ii),

$$\boldsymbol{T=2\frac{mg}{a}(a\cos\theta-a+b)+mg\cos\theta}.$$

The string will not be slack while $\theta\leqslant 90°$ but for $\theta>90°$ then if $T=0$ in any position, other than instantaneously at $\theta=180°$, then the string will be slack.

$T=0$ when

$$\frac{2}{a}(a\cos\theta-a+b)+\cos\theta=0$$

$$\boldsymbol{\cos\theta=\frac{2}{3}\left(\frac{a-b}{a}\right)}.$$

Thus the string will go slack if this value of $\cos\theta$ is consistent with $90°<\theta<180°$,

i.e. $$-1<\frac{2}{3}\left(\frac{a-b}{a}\right)<0.$$

Examining the L.H.S. of this inequality:

$$-3a<2a-2b$$

$$\boldsymbol{b<\frac{5}{2}a}$$

and the R.H.S.:

$$2(a-b)<0$$

$$\boldsymbol{a<b}.$$

Therefore,

$$\boldsymbol{a<b<\frac{5}{2}a}$$

are the conditions for the string to go slack.

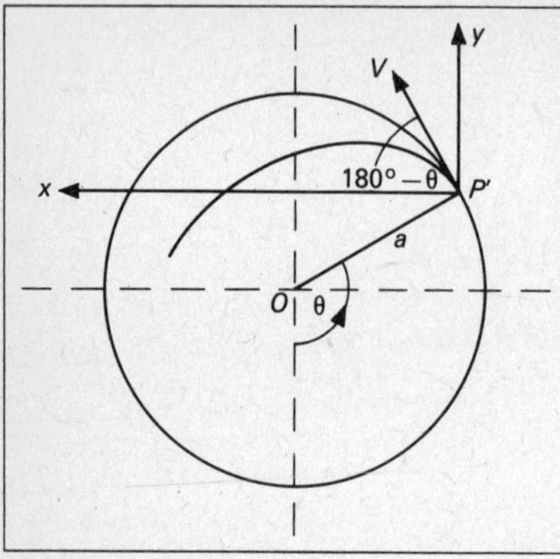

Figure 41

Refer to figure 41.

When the **string goes slack**, the particle moves as a **projectile** under the influence of gravity only. From equation (ii) above, its velocity, V, at this point is given by,

$$V^2 = 2g\left(b - a + \frac{2}{3}(a - b)\right)$$

$$\boldsymbol{V = \left(\frac{2}{3}g(b - a)\right)^{1/2}}$$

and it is **at a height**

$$a\cos(180° - \theta) = \boldsymbol{\frac{2}{3}(b - a)}$$

above the fixed point *O*.

Analysing the motion after the string becomes slack, taking the axes as shown in the diagram,

$$\ddot{y} = -g \qquad \ddot{x} = 0$$

$$\dot{y} = V\sin\theta - gt \qquad \dot{x} = V\cos\theta$$

$$y = V\sin\theta\, t - \tfrac{1}{2}gt^2 \qquad x = V\cos\theta\, t$$

The particle reaches its highest point when $\dot{y}=0$, i.e.

$$t=\frac{V\sin\theta}{g}.$$

At this time it has **travelled distance**

$$y=\frac{V^2\sin^2\theta}{g}-\frac{g}{2}\frac{V^2\sin^2\theta}{g^2}$$

$$=\frac{V^2\sin^2\theta}{2g}, \textbf{ vertically.}$$

Since $$\cos\theta=\frac{2}{3}\left(\frac{a-b}{a}\right)$$

then, $$\sin\theta=\left(1-\frac{4}{9}\left(\frac{a^2-2ab+b^2}{a^2}\right)\right)^{1/2}$$

$$=\left(\frac{5a^2+8ab-4b^2}{9a^2}\right)^{1/2}.$$

Therefore, **the greatest height reached by the particle *P* above its lowest point**, is

$$a+\frac{2}{3}(b-a)+\frac{2g}{3}\left(\frac{b-a}{2g}\right)\left(\frac{5a^2+8ab-4b^2}{9a^2}\right)$$

$$=\frac{a+2b}{3}+\frac{(b-a)(5a-2b)(a+2b)}{27a^2}$$

$$=\frac{(a+2b)(9a^2+(b-a)(5a-2b))}{27a^2}$$

$$=\frac{(a+2b)^2(4a-b)}{27a^2}.$$

Example 3

(a) A small smooth ring P of mass m is threaded on to a light inextensible string of length $2a$, whose ends are tied to fixed points A and B, where A is distant a vertically above B. Show that it is not possible for the particle to move in a horizontal circle at constant speed with AP and BP equally inclined to the vertical, but that it can move in a horizontal circle with BP horizontal, and find its speed in this case.

(b) A smooth, hollow, right circular cone with vertex downwards, is rotating with angular velocity ω about its axis which is vertical. A particle is at relative rest on the inside of the cone at a vertical height h above the vertex. Prove that

$$\omega = \sqrt{\frac{g}{h}} \cot \alpha$$

where α is the semi-vertical angle of the cone.

(S.U.J.B.)

(a) Refer to figure 42 (i).

If it were possible for the ring to move in a horizontal circle at constant speed, with AP and BP equally inclined to the vertical, then the diagram would be as shown.

The ring is smooth and the string light and inextensible, thus the tension is transmitted throughout its length.

Vertically, there should be no resultant force.

Thus, $T \cos \alpha = T \cos \alpha + mg$, which is impossible.

We therefore conclude that **this motion is not possible.**

Refer to figure 42 (ii).

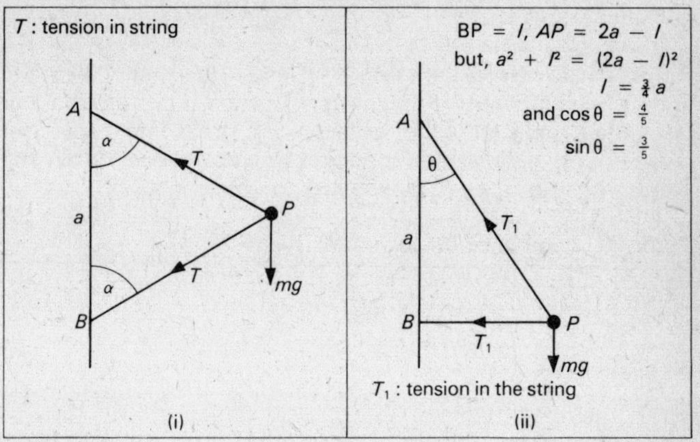

Figure 42

Let the ring, P be moving in a horizontal circle with speed V. The tension, T_1, is transmitted throughout the length of the string.

Vertically there is no resultant force.

Thus, $$T_1 \cos\theta = mg$$

$$\boldsymbol{T_1 = \frac{5}{4}mg.}$$

Horizontally, the resultant force must provide the necessary radial force,

$$\frac{mV^2}{l}$$

towards the centre, in order to keep the ring on its circular path.

Thus, radially (horizontally),

$$T_1 \sin\theta + T_1 = \frac{mV^2}{l}$$

$$\boldsymbol{V = \sqrt{\frac{3ag}{2}}.}$$

(b) Refer to figure 43.

If the particle is at relative rest inside the cone, then the horizontal component of the normal reaction, R, of the cone on the particle, must provide the necessary radial force, $m\omega^2 r$, in order to keep it moving in a horizontal circle, radius r, with the **same angular speed, ω, as that of the cone.**

Vertically, there must be no resultant force.

Thus, $$R \sin \alpha = mg.$$

Radially (horizontally),

$$R \cos \alpha = m \omega^2 h \tan \alpha.$$

Combining these equations,

$$\boldsymbol{\omega = \sqrt{\frac{g}{h}} \cot \alpha.}$$

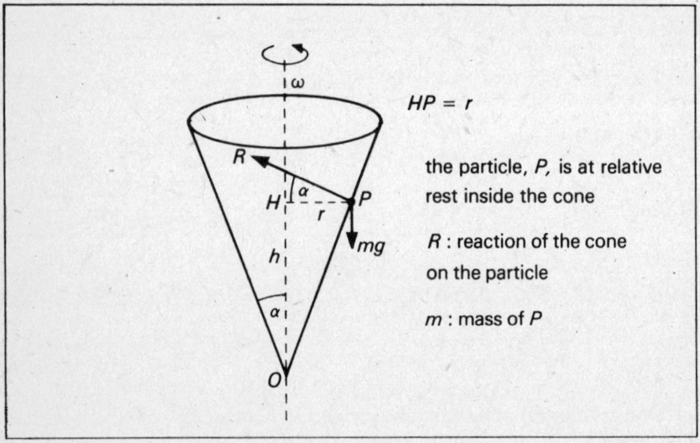

Figure 43

Chapter 10
Simple Harmonic Motion

Example 1

A particle, P, is describing simple harmonic motion in the horizontal line $ADCB$, where $AD=DC=\frac{1}{2}CB$. The speed of P as it passes through C is $5\,\text{ms}^{-1}$ and P is instantaneously at rest at A and B. Given that P performs 3 complete oscillations per second, calculate:

(i) the distance AB
(ii) the speed of P as it passes through D
(iii) the distance of P from C at an instant when the acceleration of P is $18\pi\,\text{ms}^{-1}$
(iv) the time taken by P to go directly from D to A.

(A.E.B.)

$AD = DC = \frac{1}{2}CB = y$
P moves with S.H.M. along $ADCB$

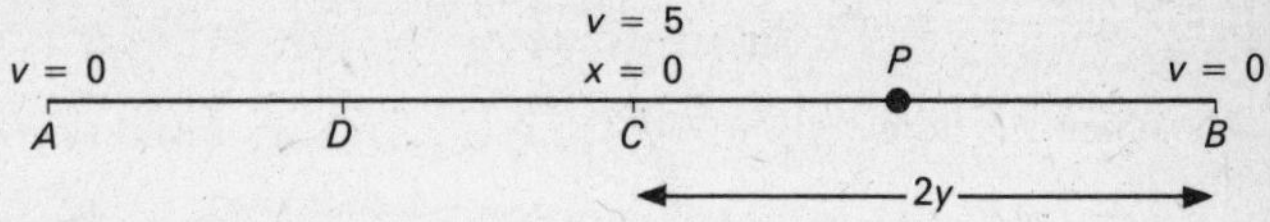

particle performs 3 complete oscillations per second
$\Longrightarrow$ period of oscillation $= \frac{1}{3}$ sec

Figure 44

Refer to figure 44.

We shall apply the standard equations of motion for S.H.M.:

$$\text{'}\frac{dv}{dt}=-\eta^2x\text{'},\quad \text{'}v^2=\eta^2(a^2-x^2)\text{'} \text{ and } \text{'}T=\frac{2\pi}{\eta}\text{'}.$$

(i) The maximum speed occurs when $x=0$ and $v=\pm\eta a$.

We are given that at $x=0$, $v=5$.

Thus, $$y=\frac{5}{2\eta}.$$

The period of oscillation, T, is given as $\frac{1}{3}$ second.

Thus, $$\frac{2\pi}{\eta}=\frac{1}{3},\quad \eta=6\pi \text{ and } y=\frac{5}{12\pi}.$$

Therefore, $$\boldsymbol{AB}=4y=\frac{\mathbf{5}}{\mathbf{3\pi}}.$$

(ii) The speed of P as it passes through D is given by the value of v when $x=-a$ in the standard S.H.M. equation.

Thus,

$$v^2=36\pi^2(4y^2-a^2)$$

$$\boldsymbol{v}=\frac{\mathbf{5\sqrt{3}}}{\mathbf{2}}\,\mathbf{ms^{-1}}.$$

(iii) When the acceleration of P is $18\pi\,\text{ms}^{-2}$, then substituting in

$$\text{'}\frac{dv}{dt}=-\eta^2x\text{'},$$

$$18\pi=-36\pi^2x,\quad \boldsymbol{x}=-\frac{\mathbf{1}}{\mathbf{2\pi}}.$$

Thus *P* is $\frac{1}{2\pi}$ m from *C*

when its acceleration is $18\pi\,\text{ms}^{-2}$.

(V.)

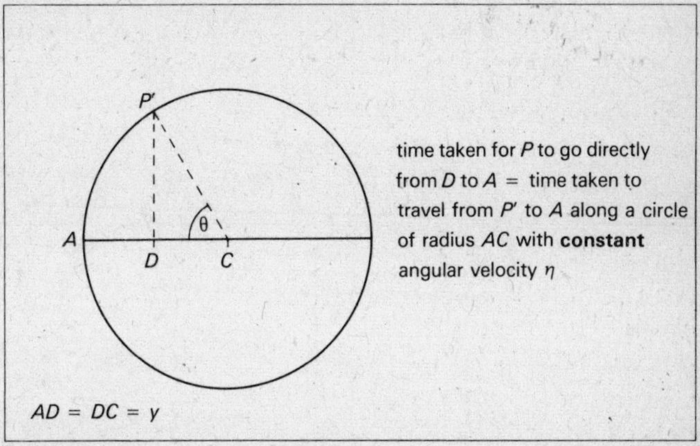

Figure 45

Refer to figure 45.

$$\cos\theta = \frac{y}{2y}, \quad \theta = \arccos\frac{1}{2} = \frac{\pi}{3}.$$

The time, t, taken to travel through angle θ along arc $P'A$ at constant angular velocity η, is given by

$$\boldsymbol{t} = \frac{\theta}{\eta} = \frac{\pi/3}{6\pi} = \frac{\mathbf{1}}{\mathbf{18}}\,\mathbf{s}.$$

Example 2

A light elastic string AB, of natural length a and modulus kmg, has a particle of mass M attached to B and A is attached to a fixed point. When the particle is hanging freely it is pulled down a distance $\frac{1}{2}a$ to C and released from rest. Show that, if $k \leqslant 2$, the particle performs complete simple harmonic oscillations. For the case $k=4$, find the total time for the particle to reach C again.

(S.U.J.B.)

Refer to figure 46 (i).

When the particle is hanging freely it is in equilibrium under the tension, T_e, and its weight mg. Thus, $T_e = mg$.

Also, by Hooke's Law,

$$T_e = \frac{\lambda e}{a} = \frac{kmge}{a}.$$

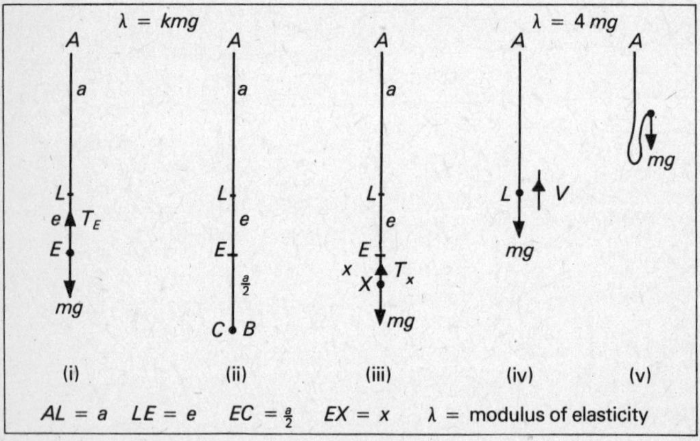

Figure 46

Combining these equations,

$$e = \frac{a}{k}.$$

The particle is pulled down distance $a/2$ from its equilibrium position, see figure 46 (ii). It is then released. We shall examine the equations of motion in the general position X, see figure 46 (iii). [The extension is measured as $(e+x)$ in anticipation of E being the centre of any S.H.M. oscillations.]

T_x is the tension in the string when $EX = x$.

Applying Newton's second law, $mg - T_x = m\ddot{x}$.

By Hooke's Law,

$$T_x = \frac{kmg(e+x)}{a}.$$

Combining these equations,

$$mg - \frac{kmge}{a} - \frac{kmgx}{a} = m\ddot{x}$$

$$\ddot{x} = \frac{-kgx}{a}.$$

This is an equation of motion for S.H.M. with **period**

$$T = 2\sqrt{\frac{a}{kg}}$$

and **centre** E, (where $x=0$),

and amplitude $\frac{a}{2}$, (where $\dot{x}=0$).

The oscillations will be **complete** provided that the string remains stretched when the particle rises to its full amplitude above E. Therefore there will be **complete S.H.M. oscillations** if

$$e = \frac{a}{k} \leqslant \frac{a}{2}, \text{ i.e. } k \leqslant 2.$$

When $k=4$, see figures 46 (iv) and (v), the particle will perform S.H.M. from C until L, whereupon the string will become slack and the particle will travel under the influence of gravity only.

Let the particle be travelling with velocity V when it reaches L and let t_1 be the time it takes to reach the greatest height it achieves above L.

From the standard equation of motion for S.H.M.,

$$\text{`}v^2 = \eta^2(a^2 - x^2)\text{'}$$

Where, in this case,

$$\text{`}\eta^2\text{'} = \frac{4g}{a}, \quad \text{`}a\text{'} = \frac{a}{2}, \quad \text{`}x\text{'} = \frac{a}{4}, \quad \text{`}v\text{'} = V,$$

then, $$V=\frac{4g}{a}\left(\frac{a^2}{4}-\frac{a^2}{16}\right).$$

$$V=\frac{1}{2}\sqrt{3ag}.$$

Applying the standard equations of motion for **constant** acceleration, '$v=u+at$', to the motion from L to the greatest height,

$$0=\frac{1}{2}\sqrt{3ag}-gt_1$$

$$t_1=\frac{1}{2}\sqrt{\frac{3a}{g}}.$$

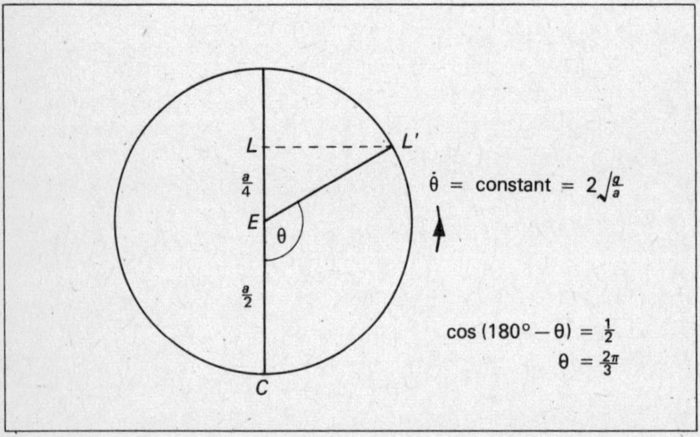

Figure 47

Refer to figure 47.

The time, t_2, the particle takes to travel from C to L is the same time as it would take the particle to travel the corresponding arc CL' with constant angular velocity,

$$\dot{\theta}=2\sqrt{\frac{g}{a}}.$$

Thus, $$t_2=\frac{2\pi/3}{2\sqrt{g/a}}=\frac{\pi}{3}\sqrt{\frac{a}{g}}.$$

Therefore **the total time taken for the particle** to travel from C to L with S.H.M., from L to its greatest height and down to L again under the influence of gravity only, and then again with S.H.M. down to C, is

$$2t_1+2t_2=\sqrt{\frac{3a}{g}}+\frac{2\pi}{3}\sqrt{\frac{a}{g}}$$

$$=\left(\sqrt{3}+\frac{2\pi}{3}\right)\sqrt{\frac{a}{g}}.$$

Example 3

A particle of mass m is attached to a fixed point A by a light elastic string of natural length $2a$ and modulus of elasticity λ. It is attached to another fixed point B by a light elastic string of natural length $5a$ and modulus of elasticity 2λ. A and B are in a smooth horizontal plane. $AB=10a$. M is the midpoint of AB, and E is the equilibrium position for the particle. Find EM.

The particle is released from rest at P, the midpoint of EM. Prove that the subsequent motion is Simple Harmonic and find the amplitude and period.

Find the speed of the particle as it passes through Q, the midpoint of EP, and show that the time taken to move from Q to E is

$$\frac{\pi}{6\omega}, \text{ where } \omega^2=\frac{9\lambda}{10am}.$$

Refer to figure 48 (i).

E is the equilibrium position, hence $T_A=T_B$.

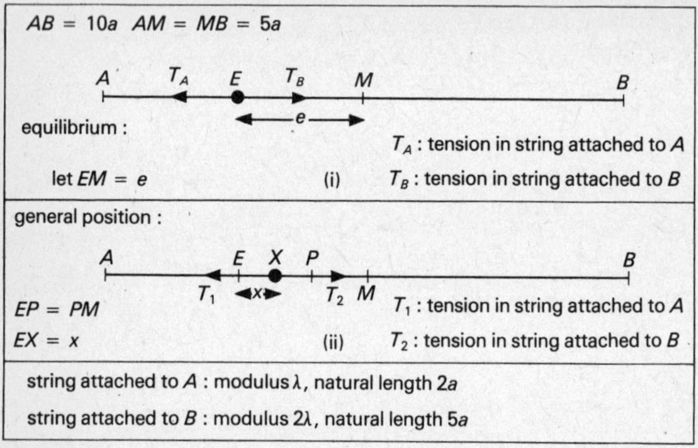

Figure 48

By Hooke's Law, $\left(`T=\dfrac{\lambda x}{l}'\right)$,

$$T_A=\frac{\lambda(3a-e)}{2a} \text{ and } T_B=\frac{2\lambda e}{5a}.$$

Equating,

$$\frac{\lambda(3a-e)}{2a}=\frac{2\lambda e}{5a}, \quad e=\frac{5}{3}a.$$

Refer to figure 48 (ii).

The particle is released from rest at P, the midpoint of EM. We shall investigate the subsequent motion by examining the forces acting at the general position X, measuring the displacement x from E, (in anticipation of the equilibrium position being the centre of S.H.M.).

Applying Newton's second law, in the direction of increasing x,

$$T_2-T_1=m\ddot{x}.$$

By Hooke's Law,

$$T_2 = \frac{2\lambda}{5a}(e-x), \quad T_1 = \frac{\lambda}{2a}(3a-e+x).$$

Substituting, $$\ddot{x} = -\left(\frac{9\lambda}{10am}\right)x.$$

This is an equation of motion for simple harmonic motion,

with period, $T = \frac{2\pi}{\omega}$, where $\omega^2 = \frac{9\lambda}{10am}$,

and centre of oscillation $x=0$.

The particle was released from rest at P,

where $$x = \frac{5a}{6},$$

hence this is the amplitude of the S.H.M. (The two strings are always stretched, thus motion is entirely S.H.M.)

When the particle passes through Q, the midpoint of EP,

$$x = \frac{5a}{12}.$$

Substituting into the standard equation for S.H.M., '$v^2 = \omega^2(a^2 - x^2)$', the velocity at Q, V_Q, is given by

$$V_Q^2 = \omega^2\left(\frac{25}{36} - \frac{25}{144}\right), \quad V_Q = \frac{5}{12}\sqrt{3}\,a\omega.$$

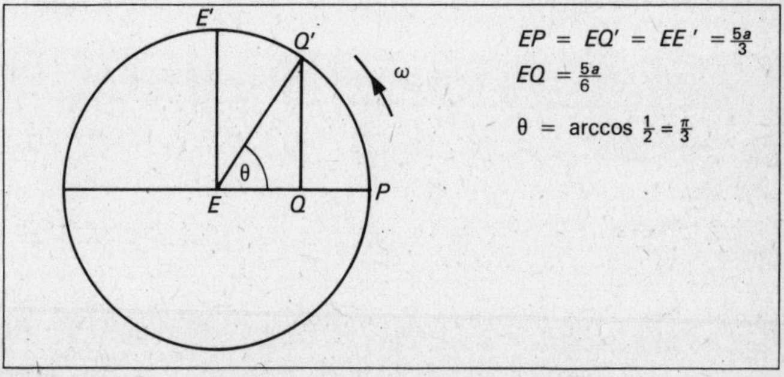

Figure 49

Refer to figure 49.

We shall inspect the uniform circular motion associated with this S.H.M.

The time taken for the particle to travel from Q to E is the same time as that which a particle would take to travel along the arc $Q'E'$ with constant angular speed ω.

The time taken to traverse an arc subtending angle θ at the centre with constant angular speed ω is

$$\frac{\theta}{\omega}.$$

Hence the time, t, taken to traverse the arc $Q'E'$ is given by,

$$t = \frac{\pi/2 - \pi/3}{\omega} = \frac{\pi}{6\omega}.$$

Chapter 11
Damped Oscillations

Example 1

A simple pendulum consists of a particle of mass m suspended from a fixed point by a light inextensible string of length l. The pendulum executes small oscillations in a fixed vertical plane and is damped by air resistance of $2mk$ times the speed of the particle, where k is a positive constant. Find the equation of motion of the particle in terms of θ, t, k, g, l, where θ is the inclination of the string to the downward vertical.

Given that $g = l(k^2 + \omega^2)$, where ω is real and that the system is released from rest where $\theta = \beta$, show that

$$\theta = \beta e^{-kt}\left(\cos \omega t + \frac{k}{\omega}\sin \omega t\right).$$

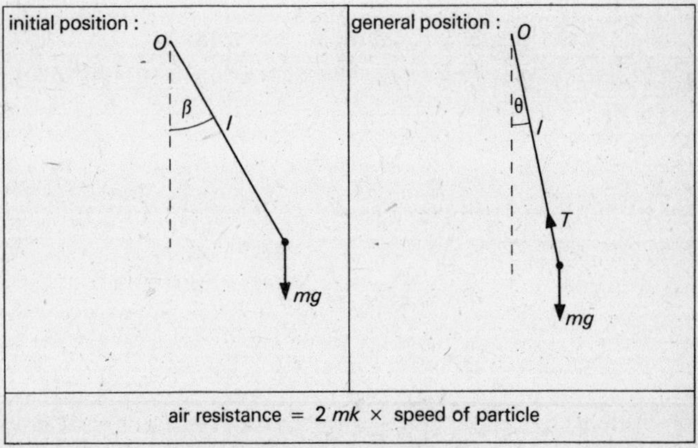

Figure 50

Refer to figure 50.

Applying Newton's second law, '$f=ma$', tangentially in the direction of increasing θ, in the general position,

$$-mg\sin\theta-2mk\,\dot{\theta}l=ml\ddot{\theta}$$

therefore $$\ddot{\theta}+2k\dot{\theta}+\frac{g}{l}\sin\theta=0.$$

For small θ, neglecting powers of θ higher than the first, $\sin\theta \triangleq \theta$, thus the required equation of motion is:

$$\boldsymbol{\ddot{\theta}+2k\dot{\theta}+\frac{g\theta}{l}=0.}$$

This is a **second order linear differential equation with constant coefficients**.

The auxiliary equation is

$$p^2+2kp+\frac{g}{l}=0$$

thus, $$p=-k\pm\sqrt{k^2-\frac{g}{l}}.$$

We are given that $g=l(k^2+\omega^2)$, therefore $p=-k\pm\sqrt{-\omega^2}$ and the roots of the auxiliary equation are **complex**.

The general form of the general solution of this equation is

$$\text{'}\theta=A'e^{(-k+i\omega)t}+B'e^{(-k-i\omega)t}\text{'}$$

Using '$e^{i\theta}=\cos\theta+i\sin\theta$' and '$e^{-i\theta}=\cos\theta-i\sin\theta$', and changing the constants A', B' we have,

$$\theta=e^{-kt}(A\cos\omega t+B\sin\omega t).$$

We are given that when $t=0$, $\theta=\beta, \dot{\theta}=0$.

Substituting,

$$\beta=A\cos 0+B\sin 0,\ \text{therefore}\ \boldsymbol{A=\beta}.$$

Differentiating and substituting,

$$\dot{\theta} = -ke^{-kt}(A\cos\omega t + B\sin\omega t) \\ + e^{-kt}(-A\omega\sin\omega t + \omega B\cos\omega t)$$

i.e. $$0 = -k\beta + B\omega$$

$$\boldsymbol{B = \frac{k\beta}{\omega}}.$$

Hence the general solution is

$$\boldsymbol{\theta = \beta e^{-kt}\left(\cos\omega t + \frac{k}{\omega}\sin\omega t\right)}.$$

Example 2

A particle, P, moves along the x-axis so that its displacement x from the origin O at the time t satisfies the differential equation

$$\frac{d^2x}{dt^2} + 4x = k,$$

where k is a constant. At times $t=0$, $\pi/4$, $\pi/2$, the values of x are 8, 9, 2, respectively. Find the value of k and find x in terms of t. Show that the greatest speed of the particle is $10\,\mathrm{ms}^{-1}$.

Sketch the graph of $\frac{dx}{dt}$ against t for $0 \leqslant t \leqslant \pi$.

If the motion starts from $t=0$ calculate when the particle is first at rest. Show that the particle is next moving in the positive direction of the x-axis with its initial speed when $t = 2.5$s.

Refer to figure 51.

The equation of motion,

$$\ddot{x} + 4x = k \qquad \text{(i)}$$

is a second order linear differential equation with constant coefficients. We shall find the **complementary function** by solving $\ddot{x} + 4x = 0$.

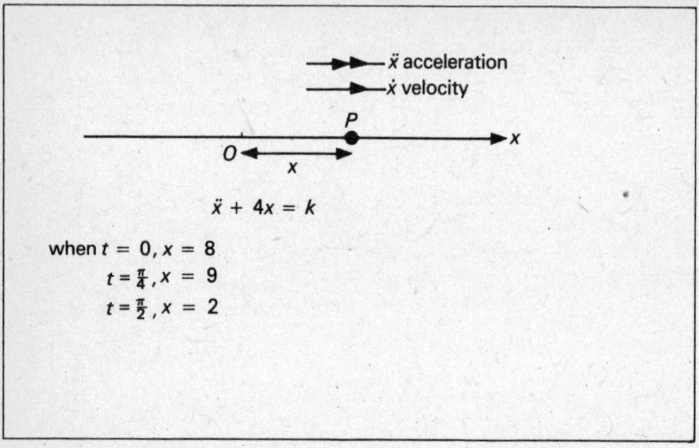

Figure 51

The **auxiliary equation** is

$$m^2+4=0$$

$$m=\pm 2i.$$

Thus the **complementary function** is $x=Ae^{2ti}+Be^{-2ti}$, where A and B are constants, or, (using the general results '$e^{i\theta}=\cos\theta+i\sin\theta$' and '$e^{-i\theta}=\cos\theta-i\sin\theta$'),

$x=A'\sin 2t+B'\cos 2t$, where A' and B' are constants.

A **particular integral**, of the same general form as the R.H.S. of the equation of motion (i) is

$$x=L \text{ where } L \text{ is a constant.}$$

Differentiating,

$$\dot{x}=0, \ddot{x}=0.$$

Substituting into equation (i)

$$0+4L=k, \quad L=\frac{k}{4}.$$

Thus the **general solution** is

$$x=A'\sin 2t+B'\cos 2t+\frac{k}{4}.$$

where A', B', k are constants.

We are given that when $t=0,\ \pi/4,\ \pi/2$ then $x=8,\ 9,\ 2$, respectively.

Substituting,

$$8=B'+\frac{k}{4}$$

$$9=A'+\frac{k}{4}$$

$$2=-B'+\frac{k}{4}.$$

Therefore $\quad k=20, \quad B'=3, \quad A'=4$

and $\quad x=4\sin 2t+3\cos 2t+5.$

The **maximum speed** occurs when $\ddot{x}=0$.

Differentiating,

$$\dot{x}=8\cos 2t-6\sin 2t$$

$$\ddot{x}=-16\sin 2t-12\cos 2t$$

Therefore, $\quad \ddot{x}=0$ when $\tan 2t=-\frac{3}{4}$.

When $\tan 2t$ is negative, $2t$ is either in the second or fourth quadrant; i.e. either $\sin 2t=\frac{3}{5}$, $\cos 2t=-\frac{4}{5}$ (second quadrant), or $\sin 2t=-\frac{3}{5}$, $\cos 2t=\frac{4}{5}$ (fourth quadrant).

In both cases the magnitude of the velocity is the same. Thus **the maximum speed** is

$$8\times\frac{4}{5}+6\times\frac{3}{5}=\mathbf{10\,ms^{-1}}.$$

In order to sketch a graph of $\dfrac{dx}{dt}$ against t, we require at least, the turning points and where it crosses the x-axis.

The **turning points** of the graph occur when $\ddot{x}=0$, i.e. when $\tan 2t=-\frac{3}{4}$, $2t=n\pi+2.5$, and $\boldsymbol{t}=\mathbf{1.25,\ 2.82,\ldots}$

From above we know that the maximum speed is $10\,ms^{-1}$.

The graph **crosses the *x*-axis** when

$$\dot{x}=0,$$

i.e. $$\tan 2t=4/3$$

$$2t=n\pi+0.927,$$

and $$\mathbf{t=0.464,\ 2.035,\ldots}$$

When

$$t=0, \quad \dot{x}=8$$

$$t=\pi/4 \quad \dot{x}=-6$$

$$t=\pi/2 \quad \dot{x}=-8$$

$$t=3\pi/4 \quad \dot{x}=-6$$

$$t=\pi \quad \dot{x}=8$$

We now have sufficient information to sketch the graph.

Refer to figure 52.

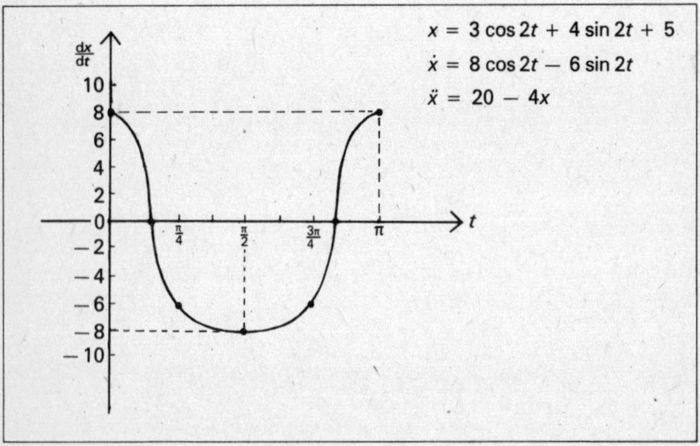

Figure 52

From above, if the motion starts at $t=0$ then the particle is first at rest when $t=0.464$ s.

The particle has a positive velocity, i.e. moving in the positive direction of the x-axis, until $t=0.464$ s. The particle then has a negative velocity until $t=2.035$ s.

It then moves in the positive direction of the x-axis and we see from the graph that it has speed, 8 units, between $t=3\pi/4$ s and $t=2.82$ s.

Substituting $\dot{x}=8$, into $\dot{x}=8\cos 2t-6\sin 2t$

$$8=8\cos 2t-6\sin 2t$$

Comparing R.H.S. with '$R\cos(2t+\alpha)$',

$$8=10\cos(2t+0.644)$$

$$\cos(2t+0.644)=0.8$$

Therefore $\quad 2t+0.644=0.644,\ 5.64,\ 6.927,$

and $\quad \boldsymbol{t=0,\ 2.5,\ldots}$

Thus **the particle is next moving in the positive direction of the x-axis with its initial speed, when $t=2.5$ s.**

Chapter 12
Forces Acting at a Point

Example 1

Given that the rigid body is in equilibrium under the action of three coplanar forces, show that the lines of action of these forces must all be parallel or all meet in a common point.

A uniform sphere of weight W and radius a, rests on a plane which is inclined at an angle α to the horizontal. When the plane is smooth, the sphere is kept in equilibrium by an inextensible string of length b, which joins a point on the surface of the sphere to a point on the inclined plane. Show that the tension in the string is

$$W \sin \alpha (a+b)(2ab+b^2)^{-1/2}.$$

When the plane is rough and the angle of inclination $\alpha, < \lambda$, the angle of friction, the sphere is kept in equilibrium by a vertical tangential force P, acting at a point on the surface of the sphere. Given that the force acts upwards in the vertical plane containing the centre of the sphere and its point of contact with the plane, find P in terms of W and α.

(A.E.B.)

If a rigid body is in equilibrium under the action of three coplanar forces, then the vector sum of the forces must be zero and their resultant turning effect must be zero.

Consider the action of three non-parallel forces $\mathbf{F}_1$, $\mathbf{F}_2$, $\mathbf{F}_3$, on a rigid body.

Refer to figure 53 (i).

The resultant of $\mathbf{F}_1$ and $\mathbf{F}_2$ passes through the point of intersection of the lines of action of those forces.

If the system is to be in equilibrium, then $\mathbf{F}_3$ must be of equal magnitude and opposite direction to $\mathbf{F}_1 + \mathbf{F}_2$ in order to reduce the vector sum to zero, and $\mathbf{F}_3$ must have the same line of action as $\mathbf{F}_1 + \mathbf{F}_2$ so that there is zero turning effect.

Therefore the forces are **concurrent**.

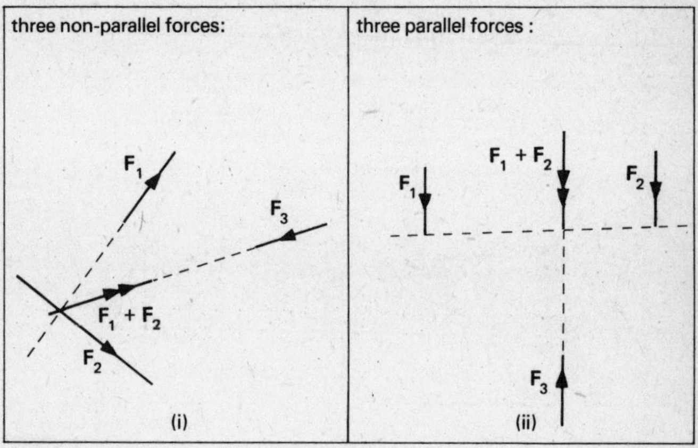

Figure 53

Refer to figure 53 (ii).

If two of the forces $\mathbf{F}_1$ and $\mathbf{F}_2$ are parallel, then the resultant of those forces is also parallel to them. If the system is to be in equilibrium, then $\mathbf{F}_3$ must be of equal magnitude to $\mathbf{F}_1 + \mathbf{F}_2$ but of the opposite direction and act along the same line of action. Consequently, $\mathbf{F}_1$, $\mathbf{F}_2$ and $\mathbf{F}_3$ must be **parallel** to one another.

Refer to figure 54 (i).

The sphere is in equilibrium under the action of the three coplanar, non-parallel, and therefore concurrent, forces: the normal reaction R, the weight of the sphere W, and the tension in the string T.

The angle between the string and the plane is β, where

$$\sin \beta = \frac{a}{a+b},$$

thus,

$$\cos\beta = \frac{((a+b)^2 - a^2)^{1/2}}{a+b}.$$

Resolving parallel to the plane:

$$T\cos\beta = W\sin\alpha$$

therefore,

$$\boldsymbol{T = W\sin\alpha\,(a+b)\,(2ab+b^2)^{-1/2}}.$$

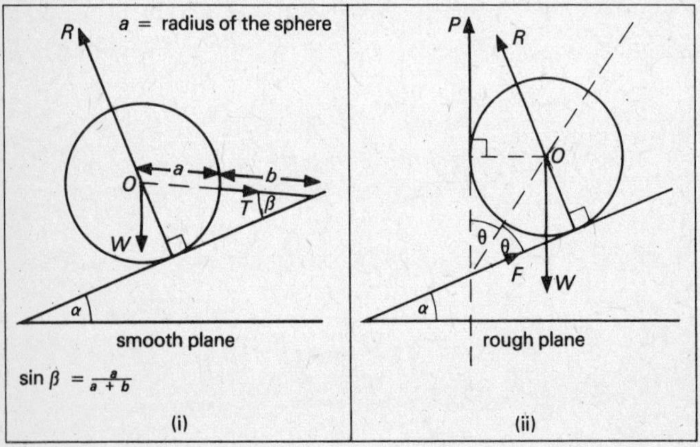

Figure 54

Refer to figure 54 (ii).

The sphere is now in equilibrium under the action of three coplanar forces: the weight, W, of the sphere; the normal reaction, R, between the plane and the sphere; and the **resultant** force of the frictional force, F, and the vertical tangential force, P.

These three forces must be concurrent, passing through the centre, O, of the sphere.

P and F are tangential forces to the sphere. Therefore, from geometrical considerations, the line passing through the point of intersection of their lines of action and the centre of the sphere, bisects the angle between their lines of action. Since this line is the line of action of the resultant of P and F, then P and F must be **equal in magnitude**.

Resolving parallel to the plane:

$$F+P\cos(90°-\alpha)=W\sin\alpha.$$

Since $F=P$ then,

$$P=\frac{W\sin\alpha}{(1+\sin\alpha)}.$$

Example 2

A block of mass 1 kg rests in equilibrium on a rough horizontal plane under the action of four horizontal forces of magnitudes and directions shown in the diagram below. The coefficient of friction between the block and the plane is 0.2.

(i) Find, by calculation or drawing, the greatest and least possible values of P when

$$\cos\alpha=\frac{\sqrt{3}}{4},\quad 0°\leqslant\alpha\leqslant 90°.$$

(ii) If α may vary in the range $0°\leqslant\alpha\leqslant 90°$, find the value of α, for which equilibrium is possible with the least value of P.

(Take $g=10\,\text{ms}^{-2}$.)

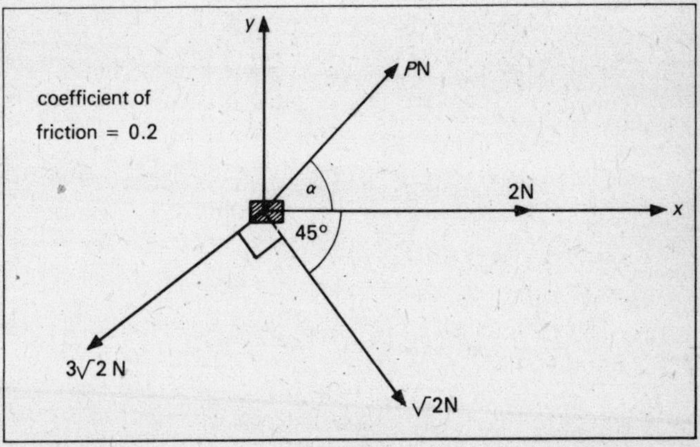

Figure 55

(i) Refer to figure 55.

Let the resultant force in the Ox direction be X and the resultant force in the Oy direction be Y.

Resolving,

$$X = 2 + \sqrt{2}\cos 45° + P\cos\alpha - 3\sqrt{2}\cos 45°$$

$$= P\cos\alpha$$

and

$$Y = P\sin\alpha - \sqrt{2}\sin 45° - 3\sqrt{2}\sin 45°$$

$$= P\sin\alpha - 4.$$

Thus the magnitude of the resultant is

$$\sqrt{P^2\cos^2\alpha + P^2\sin^2\alpha - 8P\sin\alpha + 16}$$

$$= \sqrt{P^2 - 8P\sin\alpha + 16}$$

The surface is rough and friction will be called into play in order to oppose any actual or potential motion. The block will move off in the direction of the resultant, R, of the four forces unless prevented by friction. The frictional force, F, will have the same direction as R, but the opposite sense.

The normal reaction, N, between the plane and the block, is of magnitude $1g$ N. Therefore the maximum friction that may be called into play is

$$1 \times g \times 0.2 = 2\text{ N, ('limiting friction} = \mu N\text{')}.$$

Refer to figure 56.

For equilibrium,

$$(P^2 - 2P\sqrt{13} + 16)^{1/2} = F$$

$$P^2 - 2P\sqrt{13} + 16 - F^2 = 0$$

$$\mathbf{P = \sqrt{13} \pm \sqrt{F^2 - 3}.}$$

Therefore the **greatest** and **least** values of P, for $\alpha = \arccos(\sqrt{3}/4)$, occur when the **friction, F, is limiting**, ($\sqrt{13}$ is constant and we wish to add or subtract the **maximum** value of $\sqrt{F^2-3}$) and are, respectively, $\sqrt{13}+\sqrt{4-3}$ N and $\sqrt{13}-\sqrt{4-3}$ N.

i.e. **4.6 N and 2.6 N.**

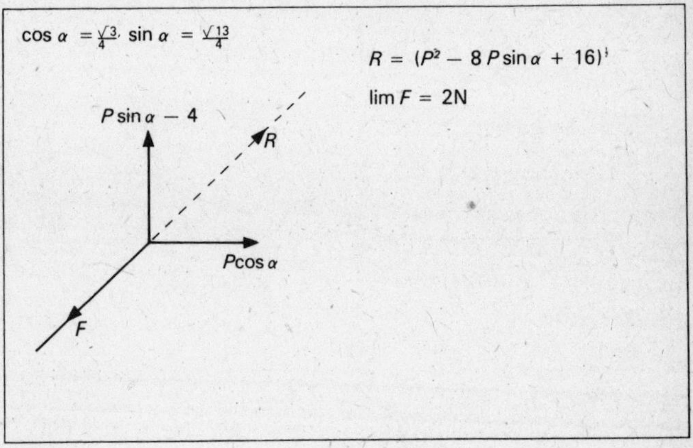

Figure 56

This part could also have been solved by **scale drawing**, see figure 57. Since the system is in **equilibrium**, the vector triangle must be **closed**.

The resultant of the forces of magnitudes $3\sqrt{2}$ N, $\sqrt{2}$ N, 2 N, is 4 N in the negative direction of the y-axis. OA represents this force. The direction of P is along AD. The frictional force may be represented by a line from O to any point **within or on** the circle OCB, the radius representing the **maximum** value of F, 2 N. Thus **the least value of P is represented by AC** and **the greatest value of P is represented by AB.**

Therefore **the greatest value of P is 4.6 N** and **the least value is 2.6 N** for $\alpha = \arccos(\sqrt{3}/4)$.

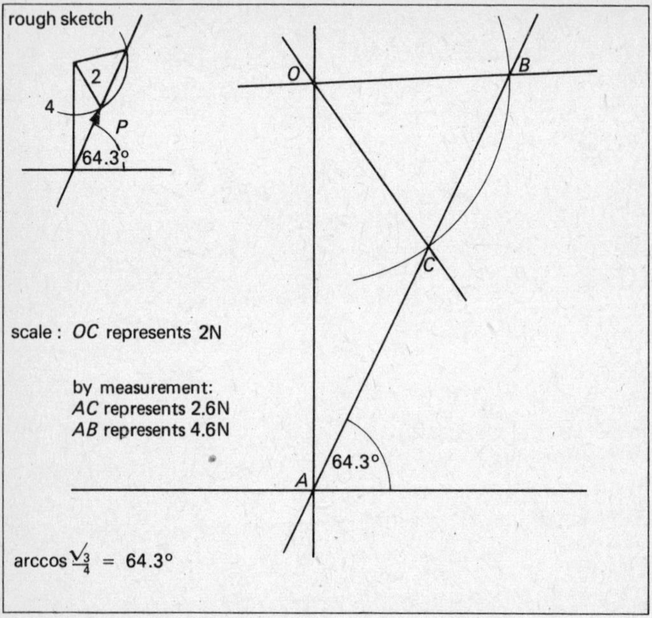

Figure 57

(ii) α may vary in the range $0° \leqslant \alpha \leqslant 90°$.

Refer to figure 58.

We are asked to find the least value of P for equilibrium. OA represents the force of magnitude 4 N (resultant of the three forces of magnitude $3\sqrt{2}$ N, $\sqrt{2}$ N and 2 N). The frictional force, F, may be represented by the line from O to a point on, or within, the circle O, radius representing 2 N, depending upon the magnitude of P and the value of α. We see that P will have its least value for any particular value of α when F has its maximum value of 2 N. (i.e. the vector triangle is OAC with the usual notation.)

Thus, we see that **the least value of P for equilibrium occurs when α=90°, and is 2 N.**

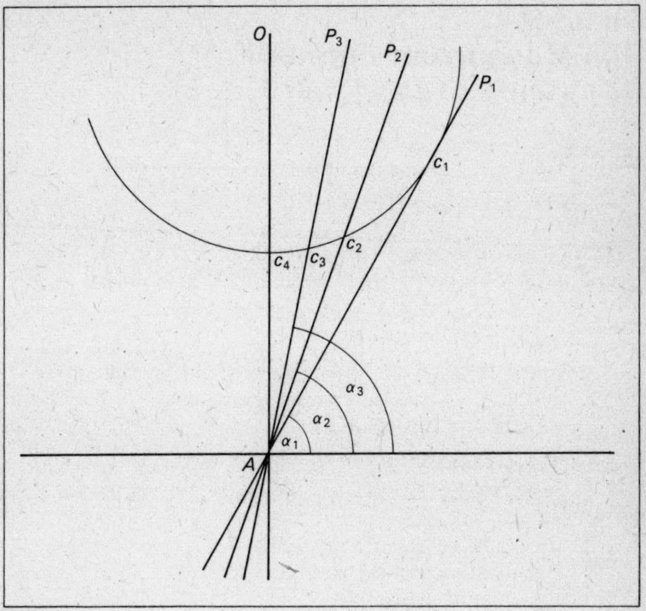

Figure 58

Chapter 13
Non-Concurrent Forces. Resultants of Coplanar Forces

Example 1

The points A, B and C have coordinates $(9a, 0)$, $(0, -4a)$, and $(6a, 4a)$, respectively, referred to the coordinate axes Ox and Oy.

(a) Forces $12P$, $15P$, $5P$ and P act along **OA**, **BC**, **CA**, and **OB**, respectively. Calculate the magnitude of the resultant of these forces and the equation of the line of action of this resultant.

(b) Forces $15P$, S and T, act along **BC**, **OA**, and **CA**, respectively. Given that these forces reduce to a couple, calculate:

(i) the values of S and T, in terms of P

(ii) the magnitude of the couple, in terms of a and P

(iii) the sense of the couple.

(A.E.B.)

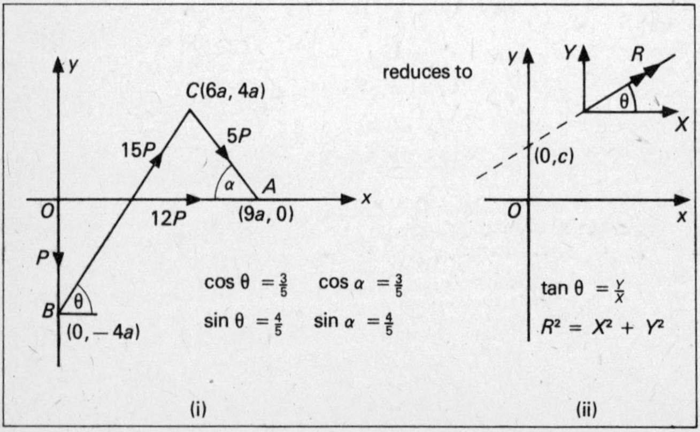

Figure 59

(a) Refer to figures 59 (i) and (ii).

Summing the components of force in the Ox, Oy directions, for the equivalent systems, and equating them,

$$X = 15P\cos\theta + 5P\cos\alpha + 12P = 24P$$

and $$Y = -P + 15P\sin\theta - 5P\sin\alpha = 7P.$$

Hence **the magnitude of the resultant force, *R*,** is

$$\sqrt{24P^2 + 7P^2} = \mathbf{25P}.$$

The equation of the line of action is of the form '$y = mx + c$', where m is the gradient and c is the intercept on the y-axis. Clearly,

$$m = \tan\theta = \frac{Y}{X} = \frac{7}{24}.$$

In order to find c, we shall take moments about O in the equivalent systems represented in figures 59 (i) and (ii), and equate them.

Taking moments about O with anticlockwise as positive:

$$4a\,15P\cos\theta - 9a\,5P\sin\alpha = -c\,24P, \quad \mathbf{c = 0}.$$

Hence **the equation of the line of action of the resultant is**

$$y = \frac{7x}{24} \text{ or } \mathbf{24y - 7x} = 0.$$

(b) Refer to figures 60 (i) and (ii).

We know that the characteristics of a system of coplanar forces reducing to a couple are that the vector sum is zero, i.e. $\Sigma X = \Sigma Y = 0$, and that the resultant moment about any point is constant and non-zero, i.e. $G = \text{const} \neq 0$.

(i) Summing in the Ox, Oy directions,

$$X = 15P\cos\theta + S + T\cos\alpha = 0$$

and $$Y = 15P\sin\theta - T\sin\alpha = 0.$$

Hence, $T = 15P$ and $S = -18P$.

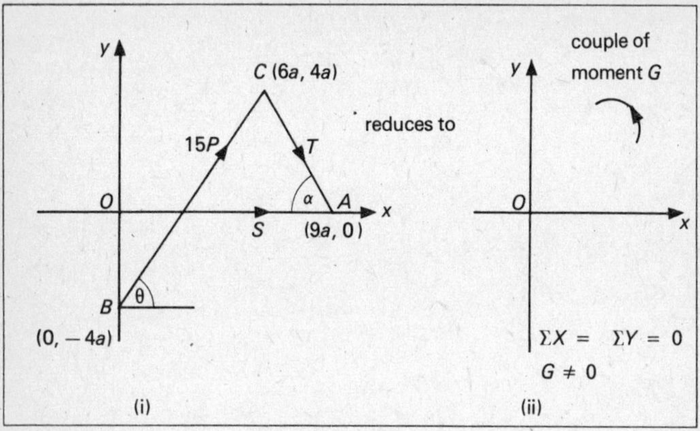

Figure 60

(ii) Taking moments about O for the equivalent systems,

$$G = 4a\,15P\cos\theta - 9a\,15P\sin\alpha = -72aP.$$

Hence **the magnitude of the couple is 72*aP*.**

(iii) **The sense of the couple is clockwise**, i.e. in the direction *ABC*.

Example 2

A non-uniform rod AB of length $6a$ rests in a horizontal position on two pegs distant a from each end. (Draw your diagram with B to the right of A). The rod will just tilt if a weight W is attached to A or a weight $2W$ is attached to B. Find the weight of the rod and the distance of its centre of gravity from A.

When the rod is resting on the pegs (with neither of the weights attached) a clockwise couple of moment Wa is applied to the rod in the vertical plane containing the rod. Find the magnitudes of the reactions of the pegs on the rod.

(S.U.J.B.)

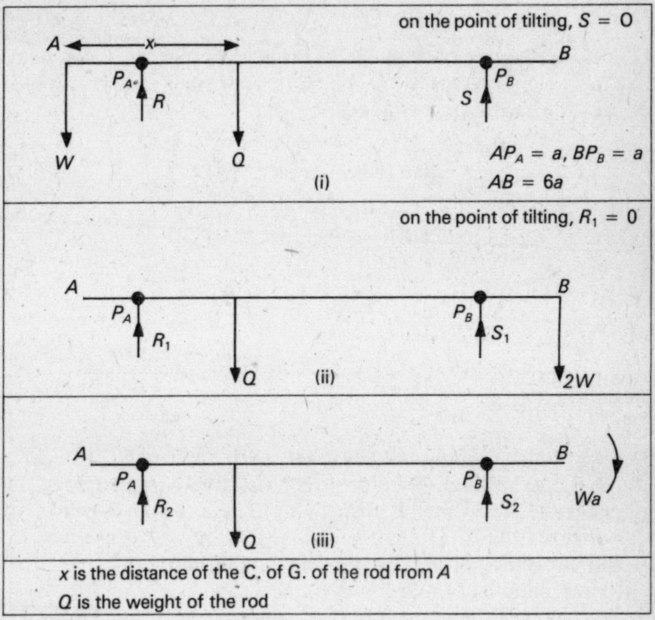

Figure 61

Refer to figure 61 (i).

The rod is on the point of tilting about P_A, therefore the reaction at P_B has just reached zero. The system is still in equilibrium. Taking moments about P_A:

$$aW = (x-a)Q. \qquad \text{(i)}$$

Refer to figure 61 (ii).

The rod is on the point of tilting about P_B, the reaction at P_A has just reached zero. Taking moments about P_B:

$$a\,2W = (5a-x)Q \qquad \text{(ii)}$$

(i) ÷ (ii), $\quad \dfrac{1}{2} = \dfrac{x-a}{5a-x}, \quad \boldsymbol{x = \dfrac{7a}{3}}.$

Substituting, $\quad \boldsymbol{Q = \dfrac{3}{4}W}.$

Refer to figure 61 (iii).

A clockwise couple of moment Wa is now applied to the system and the weights removed. A couple has constant moment about all axes. Taking moments about P_A:

$$-\frac{4}{3}a\times\frac{3}{4}W+4a\,S_2-Wa=0, \boldsymbol{S_2}=\frac{\boldsymbol{W}}{\mathbf{2}}.$$

Taking moments about P_B:

$$\frac{8}{3}a\times\frac{3}{4}W-4a\,R_2-Wa=0, \boldsymbol{R_2}=\frac{\boldsymbol{W}}{\mathbf{4}}.$$

Example 3

Two uniform rods, AB and AC, each of length $2a$ and weighing $3W$ and W respectively, are freely jointed at A and rest in a vertical plane with B and C on a rough horizontal floor. The coefficient of friction between either rod and the floor is μ. In limiting equilibrium,
(i) will slipping first occur at B, or at C?
(ii) what will be the reaction at A (giving both magnitude and direction)?
(iii) at what angle to the horizontal will the rods be inclined?

(S.U.J.B.)

Refer to figure 62.

We are given that the system is in limiting equilibrium. Hence, taking the forces as shown in figure 62(i), and considering the system as a whole:

Resolving vertically, $W+3W=R+S$.

And horizontally, $F_1=F_2$.

Taking moments about B: $3Wa\cos\theta+W\,3a\cos\theta=S\,4a\cos\theta$.

Therefore, $\boldsymbol{S}=\frac{\mathbf{3}}{\mathbf{2}}\boldsymbol{W}$ and $\boldsymbol{R}=\frac{\mathbf{5}}{\mathbf{2}}\boldsymbol{W}$.

$AB = AC = 2a$. R and S are normal reactions, F_1 and F_2 are the frictional forces. μ is the coefficient of friction between each rod and the ground.

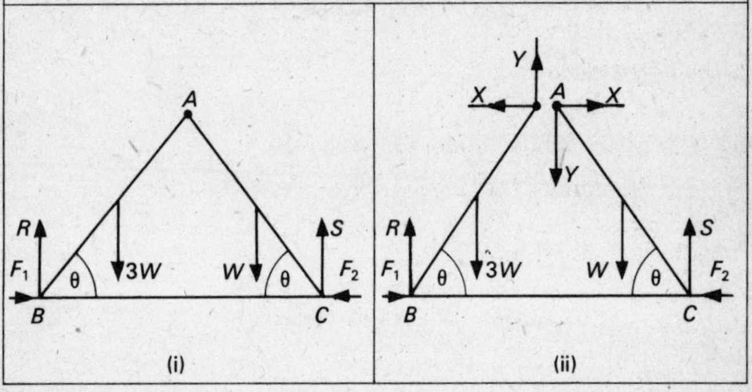

Figure 62

(i) The equilibrium is limiting, thus F_1 and/or F_2 is at its maximum value.

$$\max F_1 = \mu R = \mu \frac{5}{2} W$$

$$\max F_2 = \mu S = \mu \frac{3}{2} W.$$

Hence F_2 will reach its maximum value first and **slipping will occur at *C* first**, and **in limiting equilibrium,**

$$F_1 = F_2 = \frac{3}{2} \mu W.$$

(ii) Refer to figure 62(ii).

In order to find the reaction at A, the free joint, we separate the system at A and consider the equilibrium of the rod AB. What were **internal** forces X and Y at the joint A for the **system as a whole**, now become **external** forces when considering the **equilibrium of the rod *AB* separately**. Hence, for equilibrium of rod AB:

Resolving vertically, $Y + R = 3W$, and horizontally, $X = F_1$.

Thus,

$$X = \mu \frac{3}{2} W \text{ and } Y = \frac{W}{2}.$$

The reaction at A is therefore,

$$\sqrt{X^2+Y^2}=\frac{W}{2}\sqrt{1+9\mu^2} \text{ at } \arctan\frac{Y}{X}=\arctan\frac{1}{3\mu}$$

to the horizontal.

(iii) Taking moments about A for rod AB:

$$a\cos\theta\, 3W-2a\cos\theta\, R+2a\sin\theta\, F_1=0.$$

Substituting for R and F_1 gives:

$$\tan\theta=\frac{2}{3\mu}, \text{ and } \boldsymbol{\theta}=\mathbf{arctan}\frac{\mathbf{2}}{\mathbf{3\mu}}.$$

Example 4

ABCD is a light inextensible string of length $38a$ with $AB=15a$ and $BC=10a$. A and D are fixed to small rings, each of mass 2 kg, which can slide on a rough rail which is fixed horizontally. Masses W and W' are attached to B and C respectively, and the system rests in equilibrium with BC horizontal and distant $12a$ below AD. Show that $9W=5W'$. If the coefficient of friction between each ring and the rail is 0.25 and one of the rings is on the point of slipping, determine which it is, and the value of W.

(S.U.J.B.)

Refer to figure 63.

The system rests in equilibrium. Thus the forces in each constituent sub-system of the whole are in equilibrium. Considering equilibrium at B:

Vertically, $T_1\cos\theta=Wg$ (i)

Horizontally, $T=T_1\sin\theta$ (ii)

Considering equilibrium at C,

Vertically, $T_2\cos\alpha=W'g$ (iii)

Horizontally, $T_2\sin\alpha=T.$ (iv)

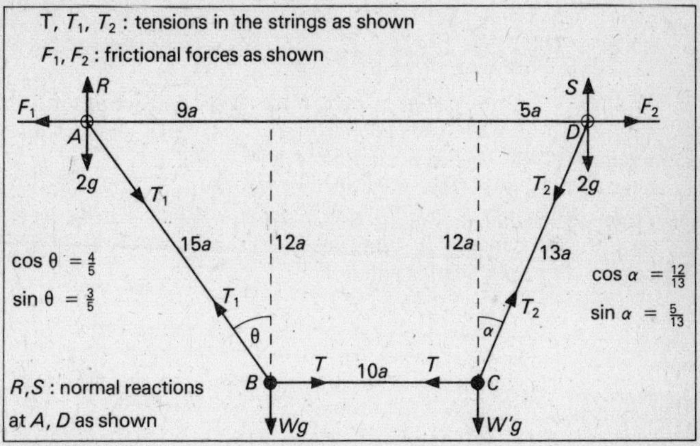

Figure 63

Combining equations (ii) and (iv), $\dfrac{T_1}{T_2}=\dfrac{25}{39}$.

Combining equations (i) and (iii), $\dfrac{13\,T_1}{15\,T_2}=\dfrac{W}{W'}$.

Therefore, $\mathbf{9W=5W'}$.

The coefficient of friction, between each ring and the rail is 0.25. Thus the maximum possible value of $F_1=0.25\,R$ and the maximum possible value of $F_2=0.25\,S$.

The system is now in **limiting** equilibrium: one of the rings is on the point of slipping. Examining the equilibrium conditions for the system as a whole: Horizontally, $F_1=F_2$.

Vertically, $R+S=4g+(W+W')g$.

Taking moments about A:

$$9a\,Wg+19a\,W'g+24a(2g-S)=0.$$

Thus, $S=(W'+2)g$ and $R=(W+2)g$.

But $W=\dfrac{5}{9}W'$ and therefore $S>R$.

Hence the **maximum value possible for the frictional force at *A* will be reached before that at *B*.** Hence **slipping will occur first at *A*.**

Example 5

Figure 64, below, shows a light framework consisting of 5 smoothly jointed rods, where AC and BD are each inclined at 45° to the horizontal. The vertical rod CD is of length 1 m and D is 1 m above the smooth horizontal plane on which A and B are standing. When a load of weight 300 newtons is hung from D, the framework is in equilibrium in a vertical plane.

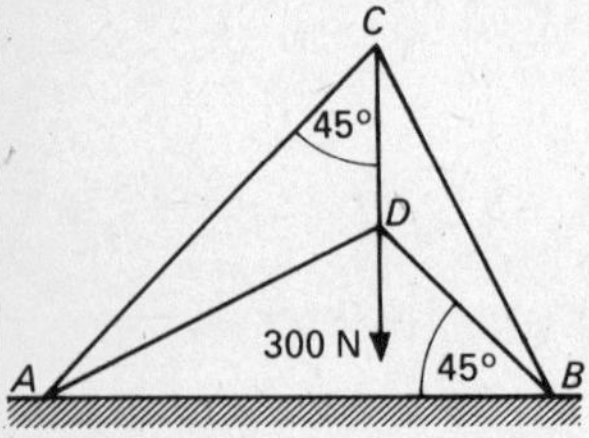

Figure 64

Calculate the magnitudes of the reactions at A and B between the framework and the plane.

Using a graphical method, or otherwise, find the magnitude of the force acting in each member of the framework and state which rods are in compression.

(A.E.B.)

Refer to figure 65.

Since the horizontal plane is **smooth**, the normal reaction, R, at A and the normal reaction, S, at B are the only **external** forces acting at these points. The only other **external** force for the system is the load of 300 N acting vertically downwards at D.

The system is in equilibrium.

Resolving vertically,

$$R + S = 300$$

Taking moments about A,

$$2 \times 300 = 3S$$

Therefore

$$\mathbf{S = 200\,N, R = 100\,N.}$$

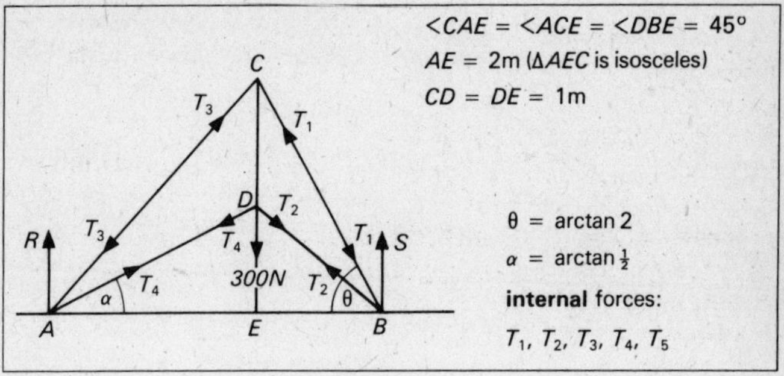

Figure 65

We shall analyse the equilibrium of the joints **separately**.

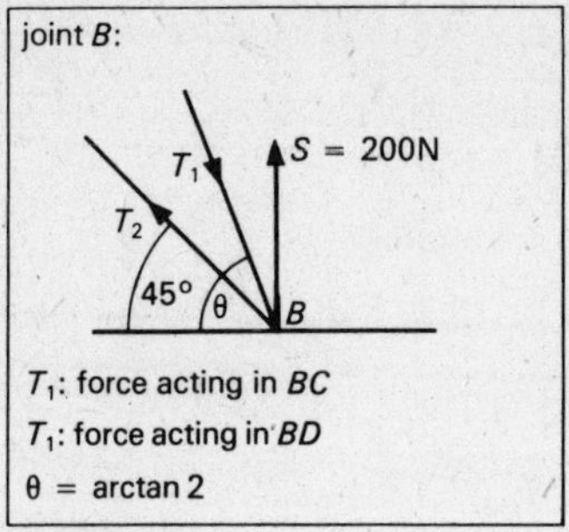

Figure 66

Joint *B*

Refer to figure 66.

Resolving vertically,

$$200 + T_2 \sin 45° = T_1 \sin \theta.$$

Resolving horizontally,

$$T_1 \cos \theta = T_2 \cos 45°.$$

Therefore, $\tan \theta = 2 = \dfrac{200 + T_2/\sqrt{2}}{T_2/\sqrt{2}}$

and $\mathbf{T_2 = 200\sqrt{2}\,N}$ **(tension in *BD*),**

$\mathbf{T_1 = 200\sqrt{5}\,N}$ **(thrust in *BC*).**

Joint *A*

Refer to figure 67.

Resolving vertically,

$$100 + T_4 \sin \alpha = T_3 \sin 45°.$$

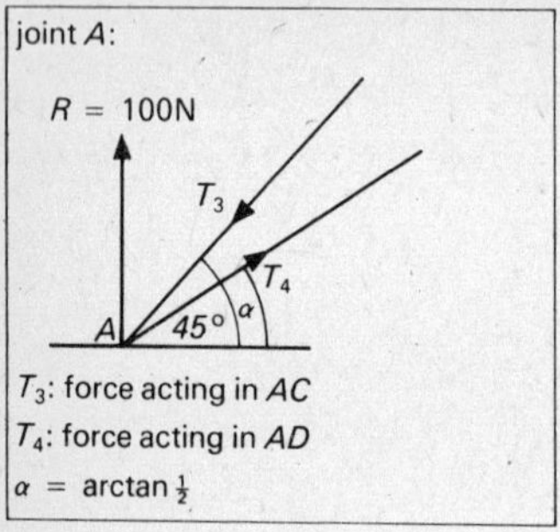

Figure 67

Resolving horizontally,

$$T_4 \cos\alpha = T_3 \cos 45°.$$

Therefore,

$$\tan\alpha = \frac{1}{2} = \frac{(T_3/\sqrt{2}) - 100}{T_3/\sqrt{2}}$$

and,

$T_3 = 200\sqrt{2}$ N (thrust in AC),

$T_4 = 100\sqrt{2}$ N (tension in AD).

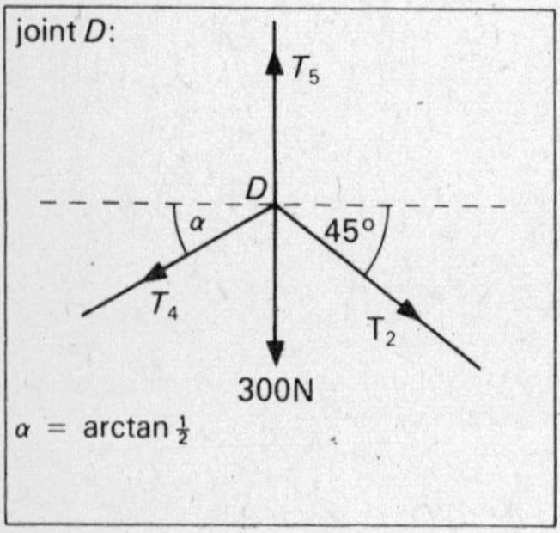

Figure 68

Joint *D*

Resolving vertically,

$$T_5 = T_4 \sin\alpha + T_2 \sin 45° + 300.$$

Resolving horizontally,

$$T_4 \cos\alpha = T_2 \cos 45°.$$

Therefore,

$$\tan\alpha = \frac{1}{2} = \frac{T_5 - 200 - 300}{200}$$

and

$$T_5 = 600 \text{ N (tension in } CD).$$

Chapter 14
Centres of Mass. Centres of Gravity

Example 1

(a) Prove that the centre of mass of a uniform lamina in the shape of the sector of a circle, radius a, subtending an angle of 2α at its centre is at a distance

$$\frac{2a\sin\alpha}{3\alpha}$$

from the centre of the circle.

(b) Find the distance from CD of the centre of mass of a uniform lamina in the shape of a square $ABCD$ of side $2a$ with a semicircular sector of radius a, diameter CD, removed from it.

(c) Find the tangent of the angle that AB would make with the downward vertical if the above lamina were freely suspended from A.

Part (a) of this question is standard bookwork and may be found by reference to any 'A' level Applied Maths text book, (e.g. *Key Facts Passbook: A-Level Applied Mathematics.*)

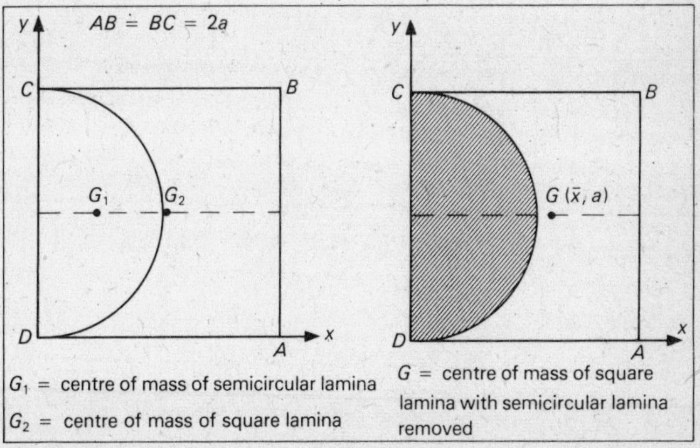

Figure 69

(b) Refer to figure 69.

The lamina may be considered to consist of the square lamina $ABCD$ minus the semicircular lamina on CD. The distance from CD of the centre of mass of the semicircular lamina is found by reference to the formula in part (a). The angle $\alpha=\pi/2$ and the radius is a. Hence the co-ordinates of G_1, the centre of mass of the semicircular lamina, are:

$$\left(\frac{4a}{3\pi}, a\right).$$

The co-ordinates of G_2, the centre of mass of the square lamina $ABCD$, are, clearly, (a, a).

Taking moments about the y-axis, with the xy plane horizontal and $\bar{x}$ as the distance of the centre of mass of the resultant lamina from CD, (m is the mass per unit area),

$$4a^2\,mg\,a-\frac{\pi a^2}{2}\,mg\,\frac{4a}{3\pi}=\bar{x}\left(4a^2-\frac{\pi a^2}{2}\right)mg$$

thus, $$\bar{\boldsymbol{x}}=\frac{\mathbf{20}\boldsymbol{a}}{\mathbf{24-3}\boldsymbol{\pi}}.$$

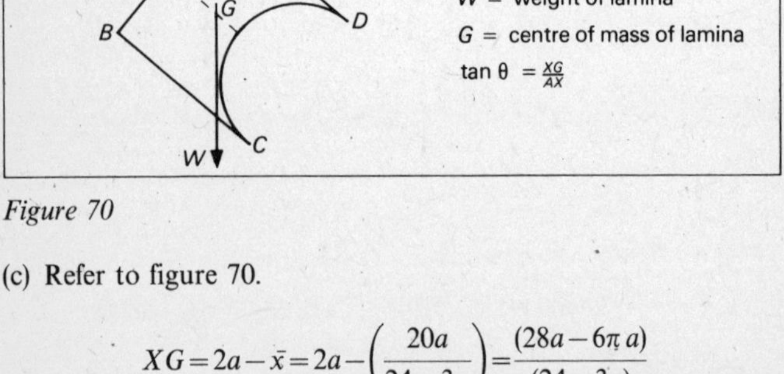

Figure 70

(c) Refer to figure 70.

$$XG=2a-\bar{x}=2a-\left(\frac{20a}{24-3\pi}\right)=\frac{(28a-6\pi a)}{(24-3\pi)}$$

and $Ax=a$.

Therefore, $$\tan\theta = \frac{XG}{AX} = \frac{28-6\pi}{24-3\pi}.$$

Example 2

Show, by integration, that the centre of gravity of a uniform lamina, in the shape of a semicircle of radius a, is situated at a distance $4a/3\pi$ from the straight edge.

Figure 71 shows a uniform lamina consisting of a semicircle $ACBD$ and an isosceles triangle OAB. The points O, C and D are collinear; $OC=3h$ and $CA=CB=CD=r$. Given that the centre of gravity of the lamina is at C, show that $9h^2=2r^2$.

The lamina, which is of mass m, has a particle of mass km attached at D. When the lamina is placed in a vertical plane with the edge OA in contact with a horizontal table, equilibrium is on the point of being broken by the lamina toppling in a vertical plane about A. Find the value of k, giving your answer to 2 significant figures.

(A.E.B.)

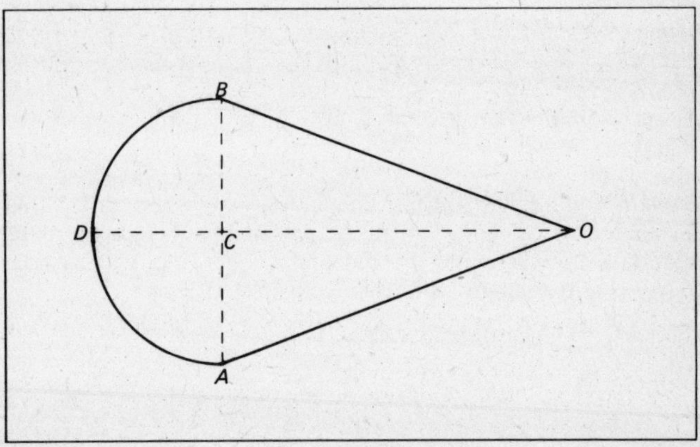

Figure 71

The first part of the question is standard book work and will not be shown here, (e.g. refer to *Key Facts Passbook: A-Level Applied Mathematics*, p.235).

The centre of gravity of the uniform lamina, shown in figure 71, is at C.

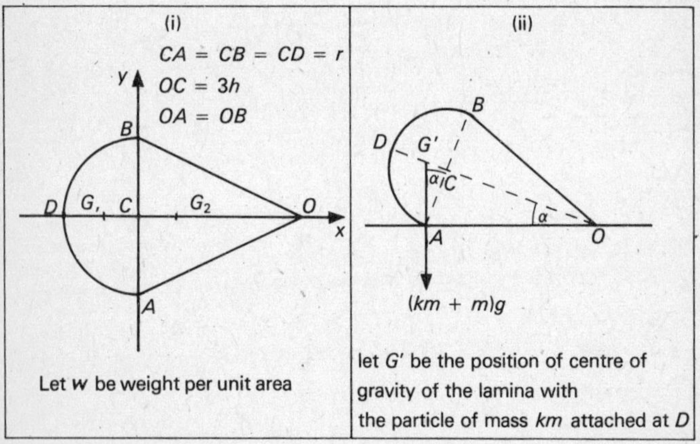

Figure 72

Refer to figure 72 (i).

The centre of gravity, G_1, of the semicircular lamina $ADCB$, is a distance $4r/3\pi$ from AB. The centre of gravity, G_2, of the triangular lamina $OBCA$ is distant $\frac{1}{3}OC = h$ from AB.

The area of the semicircular lamina is $\pi r^2/2$. The area of the triangular lamina is $\frac{1}{2} \times 2r \times 3h = 3hr$.

We shall equate the moments about the y-axis when considering the lamina to be made up of the semicircular lamina plus the triangular lamina, with the moments about the y-axis when considering the lamina as a whole, centre of gravity C.

Taking the xy plane as horizontal,

$$-h \times 3hr \times w + \frac{4r}{3\pi} \times \frac{\pi r^2}{2} \times w = 0.$$

Therefore, $\mathbf{9h^2 = 2r^2}$.

The lamina is of mass m. It now has a particle of mass km attached at D. Thus the centre of gravity, G', of the lamina with the particle attached is such that $DG':G'C=m:km$.

i.e. $$CG'=\left(\frac{k}{1+k}\right)r.$$

Refer to figure 72 (ii).

If the lamina is placed in a vertical plane with edge OA on a horizontal table and equilibrium is on the point of being broken, by toppling, then the line of action of the weight of the lamina with the particle attached, which passes through G', must also pass through A.

In triangle ACG',

$$\tan\alpha=\frac{\left(\frac{k}{1+k}\right)r}{r}$$

and in triangle OAC,

$$\tan\alpha=\frac{r}{3h}.$$

Therefore, $$\frac{k}{1+k}=\frac{r}{3h}.$$

Substituting $$9h^2=2r^2$$

then $$\frac{k}{1+k}=\frac{r}{\sqrt{2}r}$$

and $$\boldsymbol{k}=\frac{\mathbf{1}}{\sqrt{\mathbf{2}}-\mathbf{1}}=\mathbf{2.4} \text{ (correct to 2 significant figures).}$$

Chapter 15
Vector Moment. Resultant of Forces in 3 Dimensions

Example 1

A system of three forces $\mathbf{F}_1$, $\mathbf{F}_2$, $\mathbf{F}_3$, acting through points with position vectors $\mathbf{r}_1$, $\mathbf{r}_2$, $\mathbf{r}_3$, respectively, may be reduced to an equivalent force system consisting of a single force $\mathbf{F}$ through the origin and a couple of vector moment $\mathbf{G}$. If

$$\begin{aligned} \mathbf{F}_1 &= \mathbf{i}-\mathbf{j}+2\mathbf{k}, & \mathbf{r}_1 &= 3\mathbf{i}-\mathbf{j}+\mathbf{k},\\ \mathbf{F}_2 &= \mathbf{i}+3\mathbf{j}-\mathbf{k}, & \mathbf{r}_2 &= \mathbf{j}+2\mathbf{k},\\ \mathbf{F}_3 &= s\mathbf{i}+t\mathbf{j}+2\mathbf{k}, & \mathbf{r}_3 &= \mathbf{k}, \end{aligned}$$

obtain in terms of s and t, $\mathbf{F}$ and $\mathbf{G}$. Find the values of s and t when $\mathbf{G}$ and $\mathbf{F}$ are parallel to each other.

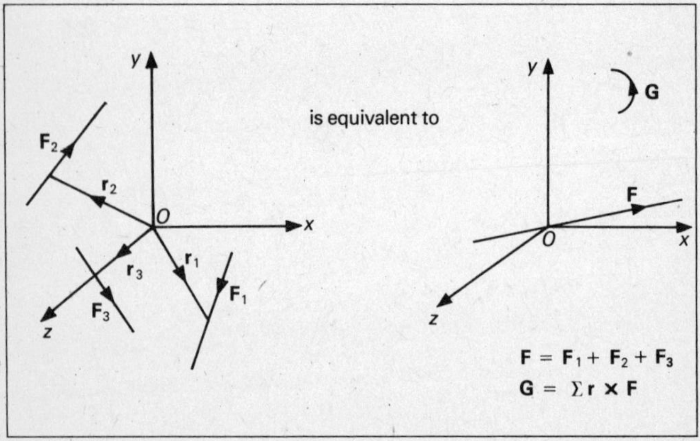

Figure 73

Refer to figure 73.

The two systems are equivalent, thus moments about any point must be identical in both systems, as must be the total components of force in any direction.

Taking vector moments about O in both systems and equating them:

$$\mathbf{G}=\mathbf{r}_1\times\mathbf{F}_1+\mathbf{r}_2\times\mathbf{F}_2+\mathbf{r}_3\times\mathbf{F}_3$$

$$=\begin{vmatrix}\mathbf{i} & \mathbf{j} & \mathbf{k}\\ 3 & -1 & 1\\ 1 & -1 & 2\end{vmatrix}+\begin{vmatrix}\mathbf{i} & \mathbf{j} & \mathbf{k}\\ 0 & 1 & 2\\ 1 & 3 & -1\end{vmatrix}+\begin{vmatrix}\mathbf{i} & \mathbf{j} & \mathbf{k}\\ 0 & 0 & 1\\ s & t & 2\end{vmatrix}$$

$$=(-8-t)\mathbf{i}+(s-3)\mathbf{j}-3\mathbf{k}.$$

Summing components of force in the **i**, **j**, **k**, directions in both systems (a couple has zero vector sum):

$$\mathbf{F}=\mathbf{F}_1+\mathbf{F}_2+\mathbf{F}_3$$

$$=(2+s)\mathbf{i}+(2+t)\mathbf{j}+3\mathbf{k}.$$

When **G** and **F** are **parallel**, their direction ratios will be equal.

Thus, when **G** is parallel to **F**,

$$(-8-t):(s-3):-3=(2+s):(2+t):3.$$

Therefore, $$\frac{-8-t}{-3}=\frac{2+s}{3},\quad \frac{s-3}{-3}=\frac{2+t}{3}$$

and $$s=3\tfrac{1}{2},\ t=-2\tfrac{1}{2}.$$

Example 2

Two forces $2\mathbf{i}+3\mathbf{j}-\mathbf{k}$ and $\mathbf{i}-2\mathbf{j}+\mathbf{k}$ act at points whose position vectors are $3\mathbf{i}+6\mathbf{j}+\mathbf{k}$ and $3\mathbf{i}-\mathbf{j}+4\mathbf{k}$ respectively. Show that the lines of action of these forces intersect and find the position vector of the point of intersection. Find also

(i) the magnitude of the resultant and a vector equation of its line of action

(ii) the vector equation of the plane containing the two forces

(iii) the moment vector of the resultant about the origin.

(A.E.B.)

(i) Refer to figure 74.

The vector equation of a line is of the form '$\mathbf{r}=\mathbf{a}+\mu\mathbf{b}$', where **a** is the position vector of a particular point on the line, μ is a constant, and **b** is a direction vector for the line.

Thus a vector equation of the line of action of $\mathbf{F}_1$ is

$$\mathbf{r}_1=(3\mathbf{i}+6\mathbf{j}+\mathbf{k})+\lambda(2\mathbf{i}+3\mathbf{j}-\mathbf{k})$$

where λ is constant.

A vector equation of the line of action of $\mathbf{F}_2$ is

$$\mathbf{r}_2=(3\mathbf{i}-\mathbf{j}+4\mathbf{k})+\mu(\mathbf{i}-2\mathbf{j}+\mathbf{k})$$

where μ is constant.

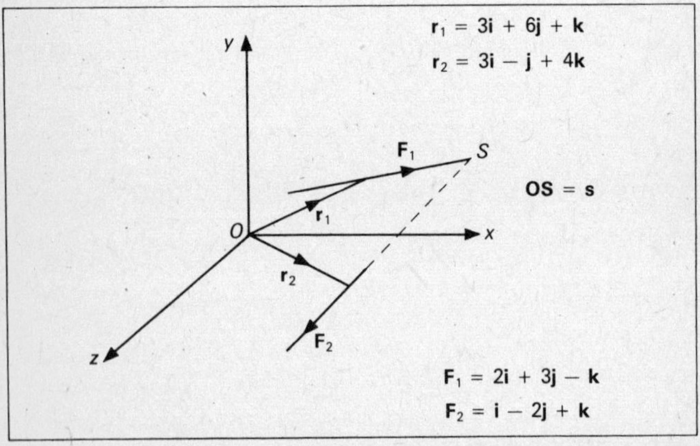

Figure 74

The line of action of the resultant force, **F**, of $\mathbf{F}_1$ and $\mathbf{F}_2$ will pass through the point of intersection, S, of the lines of action of $\mathbf{F}_1$ and $\mathbf{F}_2$. If the lines of action intersect, then there will exist particular values of λ and μ such that $\mathbf{r}_1=\mathbf{r}_2$.

i.e.

$$(3+2\lambda)\mathbf{i}+(6+3\lambda)\mathbf{j}+(1-\lambda)\mathbf{k}$$

$$=(3+\mu)\mathbf{i}+(-1-2\mu)\mathbf{j}+(4+\mu)\mathbf{k}.$$

Comparing coefficients of **i**, **j** and **k**:

$$3+2\lambda=3+\mu \qquad \text{(i)}$$

$$6+3\lambda=-1-2\mu \qquad \text{(ii)}$$

$$1-\lambda=4+\mu. \qquad \text{(iii)}$$

Combining equations (i) and (ii),

$$\boldsymbol{\mu}=-\mathbf{2} \text{ and } \boldsymbol{\lambda}=-\mathbf{1}.$$

Substituting these values in (iii),

$$1-\lambda=4+\mu$$

$$2=2$$

and **the results are consistent with intersection.**

Thus the position vector 's' of the point of intersection of the lines of action of $\mathbf{F}_1$ and $\mathbf{F}_2$ is given by,

$$\mathbf{s}=\mathbf{i}+3\mathbf{j}+2\mathbf{k}.$$

The resultant, **F**, of $\mathbf{F}_1$ and $\mathbf{F}_2$ is given by

$$\mathbf{F}=\mathbf{F}_1+\mathbf{F}_2$$

$$=3\mathbf{i}+\mathbf{j}.$$

Thus the magnitude of **F**,

$$|\mathbf{F}|=\sqrt{10}$$

and a vector equation of the line of action of **F** is

$$\mathbf{r}=\mathbf{s}+t\,\mathbf{F}, \text{ where } t \text{ is a constant}$$

$$\mathbf{r}=\mathbf{i}+3\mathbf{j}+2\mathbf{k}+t(3\mathbf{i}+\mathbf{j}).$$

(ii) Refer to figure 75.

The vector equation of a plane is of the form '$\mathbf{r}.\mathbf{n}=\mathbf{s}.\mathbf{n}$' where **r** is the position vector of any point in the plane, **n** is a vector normal to the plane and **s** is the position vector of a particular point in the plane.

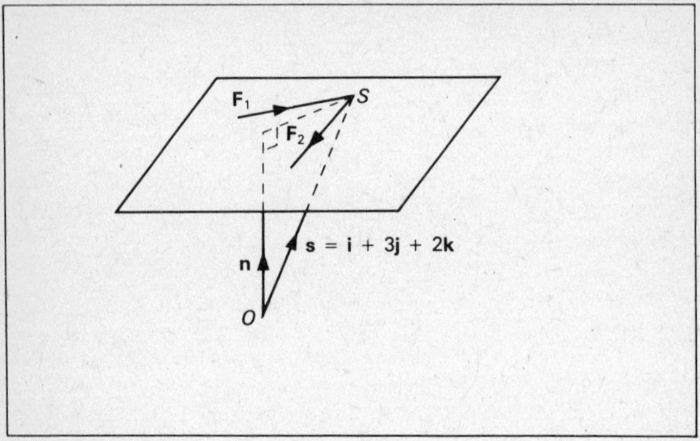

Figure 75

The vector $\mathbf{F}_1 \times \mathbf{F}_2$ is perpendicular to the plane containing $\mathbf{F}_1$ and $\mathbf{F}_2$.

$$\mathbf{F}_1 \times \mathbf{F}_2 = \begin{vmatrix} \mathbf{i} & \mathbf{j} & \mathbf{k} \\ 2 & 3 & -1 \\ 1 & -2 & 1 \end{vmatrix} = \mathbf{i}(3-2) - \mathbf{j}(2+1) + \mathbf{k}(-4-3)$$

$$= \mathbf{i} - 3\mathbf{j} - 7\mathbf{k}$$

s is the position vector of a particular point, *S*, in this plane.

Therefore, the vector equation of the plane containing $\mathbf{F}_1$ and $\mathbf{F}_2$ is given by

$$\mathbf{r}.(\mathbf{i} - 3\mathbf{j} - 7\mathbf{k}) = (\mathbf{i} + 3\mathbf{j} + 2\mathbf{k}).(\mathbf{i} - 3\mathbf{j} - 7\mathbf{k})$$

$$\mathbf{r}.(\mathbf{i} - 3\mathbf{j} - 7\mathbf{k}) = -22$$

(iii) The **moment vector**, **G**, of the resultant force about the origin is given by

$$\mathbf{G} = \mathbf{s} \times \mathbf{F} = \begin{vmatrix} \mathbf{i} & \mathbf{j} & \mathbf{k} \\ 1 & 3 & 2 \\ 3 & 1 & 0 \end{vmatrix}$$

$$= -2\mathbf{i} + 6\mathbf{j} - 8\mathbf{k}.$$

Example 3

A, *B*, *C* and *D* have position vectors $3\mathbf{i}+\mathbf{j}-4\mathbf{k}$, $-\mathbf{i}+8\mathbf{j}$, $4\mathbf{i}-\mathbf{j}-2\mathbf{k}$, and $-2\mathbf{i}+9\mathbf{j}+10\mathbf{k}$, respectively, relative to an origin *O*; **i**, **j** and **k** being mutually perpendicular unit vectors. A force of magnitude 9 newtons acts at *A* in the direction *AC* and another force acts along *AB*. If the resultant of the two forces acts along the line *AD*, find the magnitude of the force along *AB* and the magnitude of the resultant.

PQRS is a tetrahedron and forces represented completely by 2**PS**, 4**PR**, **QS** and 2**QR** act along those edges. Show that points *E* and *F*, on *PQ* and *SR*, respectively, can be chosen so that the resultant of the four forces may be represented completely by k**EF**, and give the value of k.

What are the ratios *SF*:*FR* and *PE*:*EQ*?

(S.U.J.B.)

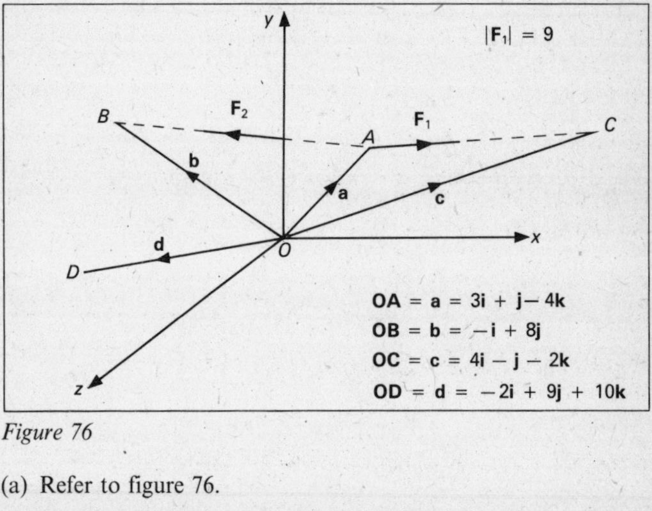

Figure 76

(a) Refer to figure 76.

Let $\mathbf{F}_1$ be the force of magnitude of 9 N acting at *A* in the direction *AC* and $\mathbf{F}_2$ be the force acting along *AB*.

A **direction vector** for $\mathbf{F}_1$ is AC.

$$\begin{aligned}\mathbf{AC} &= \mathbf{AO}+\mathbf{OC}\\ &= -\mathbf{a}+\mathbf{c}\\ &= (\mathbf{i}-2\mathbf{j}+2\mathbf{k})\end{aligned}$$

Thus, $$\mathbf{F}_1=\frac{9\,\mathbf{AC}}{|\mathbf{AC}|}=3\mathbf{i}-6\mathbf{j}+6\mathbf{k}.$$

A **direction vector** for $\mathbf{F}_2$ is AB.

$$\begin{aligned}\mathbf{AB} &= \mathbf{AO}+\mathbf{OB}\\ &= -\mathbf{a}+\mathbf{b}\\ &= -4\mathbf{i}+7\mathbf{j}+4\mathbf{k}.\end{aligned}$$

Thus, $\mathbf{F}_2=\mu(-4\mathbf{i}+7\mathbf{j}+4\mathbf{k})$ where μ is a constant.

We are told that the resultant force, $\mathbf{F}$, of $\mathbf{F}_1$ and $\mathbf{F}_2$ acts along AD. Thus a **direction vector** for $\mathbf{F}$ is $\mathbf{AD}$.

$$\begin{aligned}\mathbf{AD} &= \mathbf{AO}+\mathbf{OD}\\ &= -\mathbf{a}+\mathbf{d}\\ &= -5\mathbf{i}+8\mathbf{j}+14\mathbf{k}.\end{aligned}$$

Also, $$\begin{aligned}\mathbf{F} &= \mathbf{F}_1+\mathbf{F}_2\\ &= (3-4\mu)\mathbf{i}+(7\mu-6)\mathbf{j}+(4\mu+6)\mathbf{k}.\end{aligned}$$

The **direction ratios** of $\mathbf{AD}$ and $\mathbf{F}$ must be the same.

Therefore, $$\frac{3-4\mu}{7\mu-6}=-\frac{5}{8} \qquad \text{(i)}$$

and $$\frac{7\mu-6}{4\mu+6}=\frac{8}{14}. \qquad \text{(ii)}$$

From (i), $\boldsymbol{\mu}=\mathbf{2}$. Checking with (ii), $\boldsymbol{\mu}=\mathbf{2}$.

Hence, $$\mathbf{F}_2=-8\mathbf{i}+14\mathbf{j}+8\mathbf{k}$$

and $$|\mathbf{F}_2|=(64+196+64)^{1/2}=\mathbf{18\,N}$$

Also, $$\mathbf{F}=-5\mathbf{i}+8\mathbf{j}+14\mathbf{k}$$

and $$|\mathbf{F}|=(25+64+196)^{1/2}=\mathbf{16.9}\,\mathrm{N}.$$

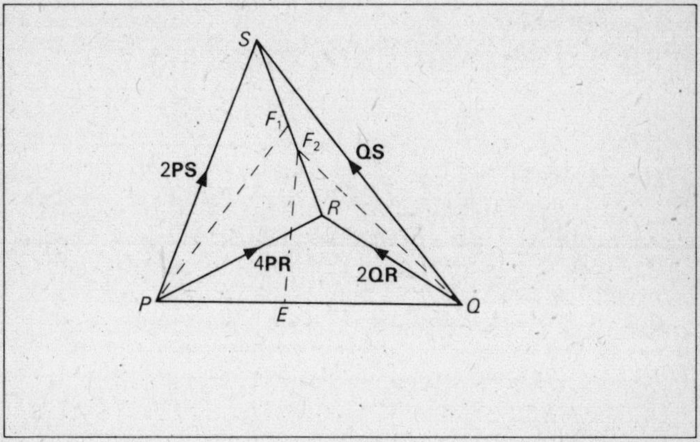

Figure 77

(b) Refer to figure 77.

By the **resultant vector theorem**, the resultant vector of the two forces represented completely by λ**OA** and μ**OB** is represented completely by $(\lambda+\mu)$**OC**, where C divides AB in the ratio $\mu:\lambda$.

Hence the resultant force of forces represented completely by 2**PS** and 4**PR** is represented completely by 6**PF**, where F_1 **divides** SR **in the ratio 2:1**.

Similarly, the resultant force of forces **QS** and 2**QR** is represented completely by 3$\mathbf{QF}_2$ where F_2 **divides** SR **in the ratio 2:1**.

Thus F_1 and F_2 are coincident. We shall call this point F. The resultant force of 6**PF** and 3**QF** is 9**EF** where E **divides** PQ **in the ratio 1:2**.

Hence **the resultant of the four forces represented completely by** 2**PS**, 4**PR**, **QS** and 2**QR** **may be represented completely by** 9**EF**, where $SF:FR=2:1$ and $PE:EQ=1:2$.

Therefore, $\boldsymbol{k}=\mathbf{9}$.

Chapter 16
Bending Moments and Shearing Force

Example 1

A uniform rod AB of length $4a$ and weight w per unit length is clamped at A in a horizontal position and a weight wa is attached to the rod at the point distant a from A. Find the vertical force and couple exerted on the rod by the clamp. P is a point of the rod distant x from A. By considering the equilibrium of the portion AP of the rod, find expressions, in terms of x, for the shearing force S and bending moment M at P. Draw sketches of the graphs of M and S against x. Find (by calculation) the point of the rod at which the bending moment is $\frac{1}{8}$th of its value at A.

(S.U.J.B.)

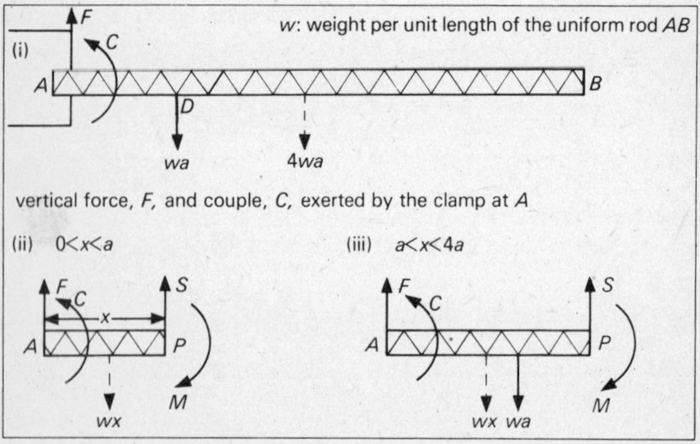

Figure 78

Refer to figure 78(i).

Consider the equilibrium of the beam.

Resolving vertically, $\qquad F = 5wa.$

Taking moments about A,

$$C = awa + 2a\,4wa = \mathbf{9a^2w}.$$

P is a point of the rod distant x from A. By considering the equilibrium of the portion AP of the rod, we shall find expressions for the shearing force, S, and the bending moment, M, at P. However, there is a concentrated force, wa, at D, and hence a discontinuity. It is therefore necessary to consider separately the sections for which $0 < x < a$ and $a < x < 4a$.

Refer to figure 78 (ii).

Resolving vertically,

$$F + S = wx$$

$$\mathbf{S = w(x - 5a).}$$

Taking moments about P,

$$C + \frac{wx^2}{2} = xF + M$$

$$\mathbf{M = \frac{w}{2}(x^2 - 10ax + 18a^2).}$$

Refer to figure 78 (iii).

Resolving vertically,

$$F + S = wa + wx$$

$$\mathbf{S = w(x - 4a).}$$

Taking moments about P,

$$C + \frac{wx^2}{2} + wa(x - a) = xF + M$$

$$M = \frac{w}{2}(x^2 + 16a^2 - 8ax).$$

Therefore, $$\mathbf{M = \frac{w}{2}(x - 4a)^2.}$$

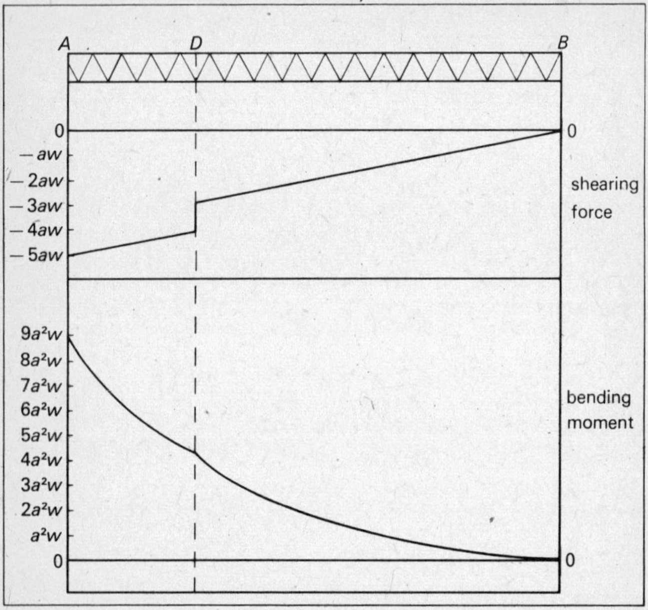

Figure 79

Refer to figure 79 for the sketches of the graphs of M and S against x.

The bending moment will be $\frac{1}{8}$th of its value at A,

when $$M = \frac{9a^2w}{8}.$$

Substituting into $M = \frac{w}{2}(x-4a)^2$,

(we see from the graph that when $M = \frac{9a^2w}{8}$, then $a < x < 4a$)

$$\frac{9a^2w}{8} = \frac{w}{2}(x-4a)^2.$$

Thus, $$(x-11a)(2x-5a) = 0$$

and $$\boldsymbol{x} = \frac{\mathbf{5a}}{\mathbf{2}} \ (x = 11a \text{ is impossible}).$$

Example 2

A light beam *ABCD* of length 20 metres rests horizontally on supports at *A* and *C*, where $AB = 10$ metres and $BC = 5$ metres. The beam carries a load of 20 newtons per metre between *B* and *D* and concentrated loads 300 newtons and 600 newtons at *A* and *B* respectively.

Show that the reactions at the supports are 500 newtons and 600 newtons. Draw the shearing force and bending moment diagrams for the beam. Find the magnitude of the maximum and minimum bending moments and state where these values occur.

(A.E.B.)

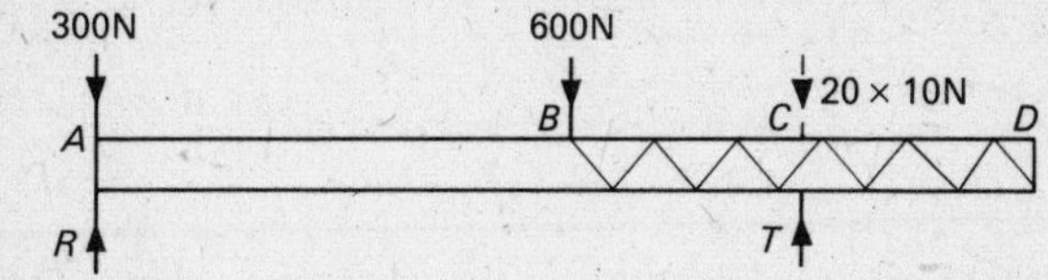

$AB = 10$m $BC = CD = 5$m

concentrated loads 300N at *A* and 600N at *B*

R : reaction at support at *A*

T : reaction at support at *C*

shows **distributed** force of 20N/m along *BD*

20 × 10N at *C* shows **overall** force from distributed force

Figure 80

Refer to figure 80.

Consider the system as a whole. Resolving vertically,

$$R + T = 300 + 600 + 200 = 1100.$$

Taking moments about *C*,

$$5 \times 600 + 15 \times 300 = 15 \times R$$

$$\mathbf{R = 500\,N,\ T = 600\,N.}$$

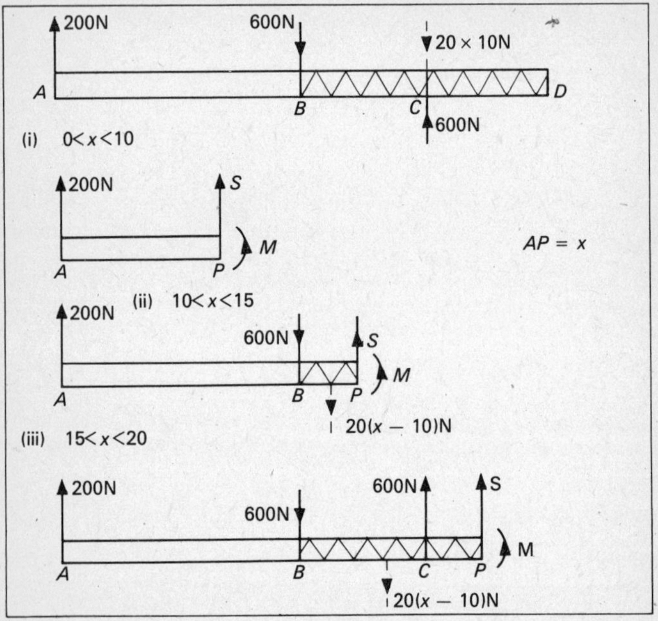

Figure 81

Let P be any point on the beam such that $AP = x$ m. Let S be the shearing force and M be the bending moment at P.

Refer to figure 81.

There are concentrated forces, and hence discontinuities, in M and S, at B and C. Thus we shall consider separately the section $0<x<10$, $10<x<15$, $15<x<20$.

(i) $0<x<10$

Resolving vertically,

$$200 + S = 0$$

$$\mathbf{S = -200\ N.}$$

Taking moments about P,

$$\mathbf{M = 200x\ Nm.}$$

(ii) $10<x<15$

Resolving vertically,

$$200-600-20(x-10)+S=0$$
$$\mathbf{S=200+20x}.$$

Taking moments about P,

$$M+(x-10)600+20(x-10)\frac{1}{2}(x-10)-200x=0$$
$$\mathbf{M=-10(x^2+20x-500)}.$$

(iii) $15<x<20$

Resolving vertically,

$$200-600-20(x-10)+600+S=0$$
$$\mathbf{S=20(x-20)}.$$

Taking moments about P,

$$M+\frac{1}{2}(x-10)20(x-10)+600(x-10)-200x-600(x-15)=0$$
$$\mathbf{M=-10(x-20)^2}.$$

Refer to figure 82 for the shearing force and bending moments diagrams.

The **greatest bending moment** (i.e. the greatest magnitude of M regardless of sign) **occurs at *B*** and is of magnitude **2000 Nm**. The **least bending moment** (i.e. the least magnitude of M regardless of sign) is **ON** and **occurs at *A*, *D*, and *E***, where E is between B and C, (from the bending moments diagram).

Setting $M=0$ in $M=-10(x^2+20x-500)$

$$x=\frac{-20\pm\sqrt{2400}}{2}\quad \text{(The negative root is impossible).}$$

Therefore, $\qquad AE=14.495\,\text{m}.$

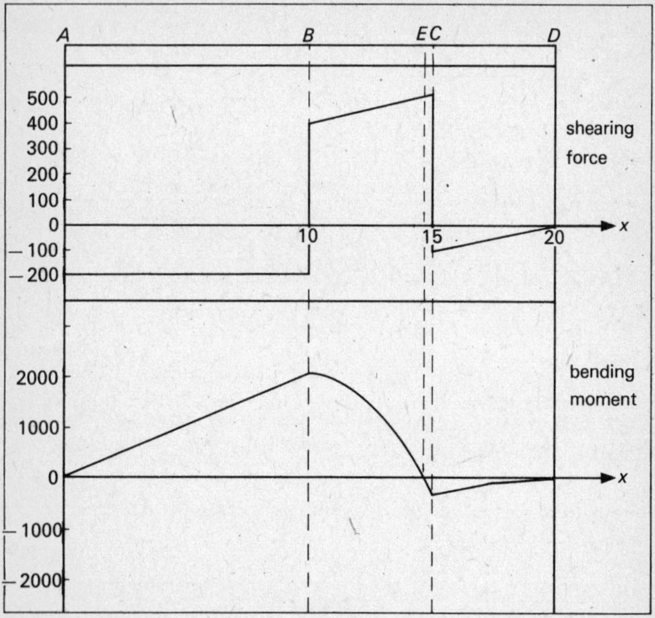

A
B
EC
D
500
400
300
200
100
0
− 100
− 200
shearing force
10
15
20
x
2000
1000
0
− 1000
− 2000
bending moment
x

Figure 82

Chapter 17
Stability of Equilibrium

Example 1

A uniform rod of length $2a$ and mass m rests with one end on a smooth vertical wall and is supported by a smooth peg distance c from the wall, the rod and the peg being in the same vertical plane perpendicular to the wall. A particle of mass $\frac{1}{2}m$ is attached to the other end of the rod with a light string. By considering the potential energy of the system show that if $3c > 4a$ there is no equilibrium position, and that if $3c < 4a$ there is one equilibrium position which is unstable.

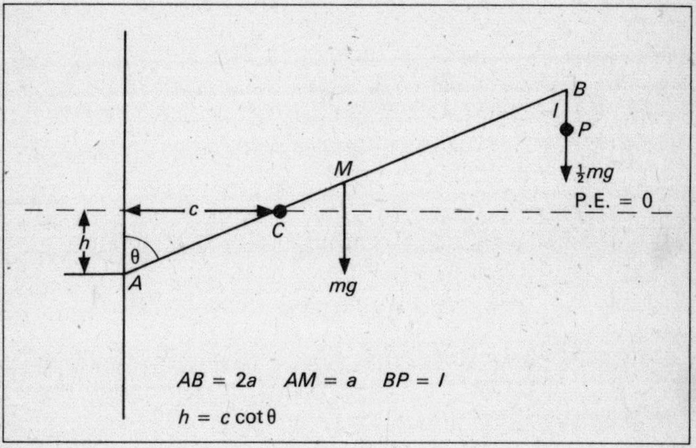

Figure 83

Refer to figure 83.

We shall take the zero potential energy line as the horizontal through the fixed peg C (the P.E.$=0$ line must not vary as θ varies).

Thus,

$$\text{P.E.} = mg(AM\cos\theta - h) + \frac{mg}{2}(BC\cos\theta - l)$$

$$= mg(a\cos\theta - c\cot\theta) + \frac{mg}{2}(2a\cos\theta - c\cot\theta - l)$$

$$= mg\left(2a\cos\theta - \frac{3c}{2}\cot\theta\right) + \text{constant.}$$

Differentiating w.r.t. θ,

$$\frac{\mathrm{d(P.E.)}}{\mathrm{d}\theta} = mg\left(-2a\sin\theta + \frac{3c}{2}\operatorname{cosec}^2\theta\right).$$

For equilibrium, $\frac{\mathrm{d(P.E.)}}{\mathrm{d}\theta} = 0$, hence equilibrium occurs when,

$$2a\sin\theta = \frac{3c}{2}\operatorname{cosec}^2\theta$$

$$\sin^3\theta = \frac{3c}{4a}, \quad \sin\theta = \left(\frac{3c}{4a}\right)^{1/3}.$$

Since $0 \leqslant \sin\theta \leqslant 1$, then equilibrium is not possible for $3c > 4a$ and there is only one equilibrium position for $3c < 4a$ at

$$\theta = \arcsin\left(\frac{3c}{4a}\right)^{1/3}.$$

We now examine the stability of this equilibrium position.

$$\frac{\mathrm{d}^2\mathrm{(P.E.)}}{\mathrm{d}\theta^2} = mg(-2a\cos\theta - 3c\operatorname{cosec}^2\theta\cot\theta)$$

$$= mg\left(-2a\cos\theta - \frac{3c\cos\theta}{\sin^3\theta}\right)$$

$$= -mg\cos\theta(2a + 4a)$$

$$= -6a\,mg\cos\theta.$$

At equilibrium, θ is clearly an acute angle and $\cos\theta > 0$,

thus, $$\frac{\mathrm{d}^2\mathrm{(P.E.)}}{\mathrm{d}\theta^2} < 0$$

and therefore we conclude that **equilibrium is unstable**.

Example 2

A and *B* are the ends of a smooth wire in the form of a semicircle of radius *a*. The wire is fixed in a vertical plane and hangs below *AB*, which is horizontal. A bead of mass *M* slides on the wire and is attached to two light strings, each of length $2a$, which pass over smooth pegs at *A* and *B*. To the other ends of the strings are attached particles of mass *m* which hang vertically under gravity. Show that if the radius to the bead of mass *M* makes an angle θ with the vertical, the potential energy of the system is

$$V_o - Mga\cos\theta + mga\,2\sqrt{2}\cos(\theta/2)$$

where V_o is a constant.

Hence deduce that the equilibrium position, in which the bead of mass *M* is at the lowest point of the wire, is stable if $m < M\sqrt{2}$.

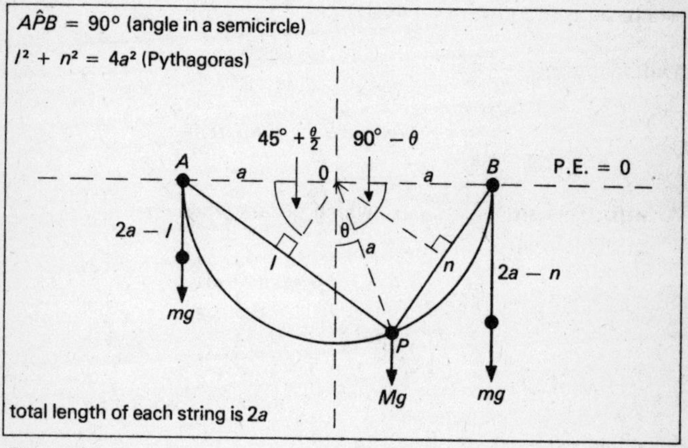

Figure 84

Refer to figure 84.

Taking the lengths of the portions of the string as shown, the potential energy of the system is given by

$$\text{P.E.} = -(2a-l)mg - (2a-n)mg - a\cos\theta\,Mg.$$

But $n = 2a\sin\left(45° - \frac{\theta}{2}\right)$ and $l = 2a\sin\left(45° + \frac{\theta}{2}\right)$.

Thus,

$$\text{P.E.} = -\left[2a - 2a\sin\left(45° + \frac{\theta}{2}\right)\right]mg$$

$$-\left[2a - 2a\sin\left(45° - \frac{\theta}{2}\right)\right]mg - a\cos\theta\, Mg$$

$$= -4amg + 2amg\left[\sin\left(45° + \frac{\theta}{2}\right) + \sin\left(45° - \frac{\theta}{2}\right)\right]$$

$$-a\cos\theta\, Mg$$

$$= -4amg - a\cos\theta\, Mg + 2\sqrt{2}\cos\frac{\theta}{2}\, amg$$

$$= \boldsymbol{V_o - Mga\cos\theta + mga\, 2\sqrt{2}\cos\frac{\theta}{2}}.$$

where V_o is a constant $(-4amg)$.

Differentiating,

$$\frac{d(\text{P.E.})}{d\theta} = Mga\sin\theta - \sqrt{2}\, mga\sin\frac{\theta}{2}.$$

At **equilibrium,** $\frac{d(\text{P.E.})}{d\theta} = 0$, which occurs when

$$Mga\sin\theta - \sqrt{2}\, mga\sin\frac{\theta}{2} = 0$$

$$\sin\frac{\theta}{2}\left(2\cos\frac{\theta}{2}M - \sqrt{2}\, m\right) = 0$$

i.e. either $\sin\frac{\theta}{2} = 0$ and $\boldsymbol{\theta = 0°}$ or $\boldsymbol{\cos\frac{\theta}{2} = \frac{m}{\sqrt{2}M}}$.

The equilibrium position, with the bead of mass M at the bottom of the wire, occurs when $\theta = 0°$.

This will be **stable** if $\boldsymbol{\frac{d^2(\text{P.E.})}{d\theta^2} > 0}$.

Differentiating again,

$$\frac{d^2(\text{P.E.})}{d\theta^2} = a\cos\theta\, Mg - \frac{\sqrt{2}}{2}\, amg\cos\frac{\theta}{2}.$$

When $\boldsymbol{\theta = 0°}$, $$\frac{d^2(\text{P.E.})}{d\theta^2} = aMg - \frac{amg}{\sqrt{2}}$$

which is **positive**, (and thus the P.E. is **minimum** and the equilibrium **stable**), when

$$aMg - \frac{amg}{\sqrt{2}} > 0$$

i.e. **when** $\boldsymbol{M\sqrt{2} > m}$.

Chapter 18
Rotation of a Rigid Body about a Fixed Axis. Calculation of Moments of Inertia

Example 1

A heavy uniform rod of length $2a$ and mass m is hinged at one end and free to rotate under gravity in a vertical plane. A particle of mass m is attached to the other end. The rod is then displaced through an angle α from its position of stable equilibrium and released from rest. Find $\dot{\theta}$ where $\theta(<\alpha)$ is the angle made by the rod with the downward vertical. Show that the force on the hinge when the rod is passing through its vertical position is

$$\frac{mg}{8}(43-27\cos\alpha)$$

vertically and 0 horizontally.

Find the period of small oscillations of the rod about its equilibrium position when α is small.

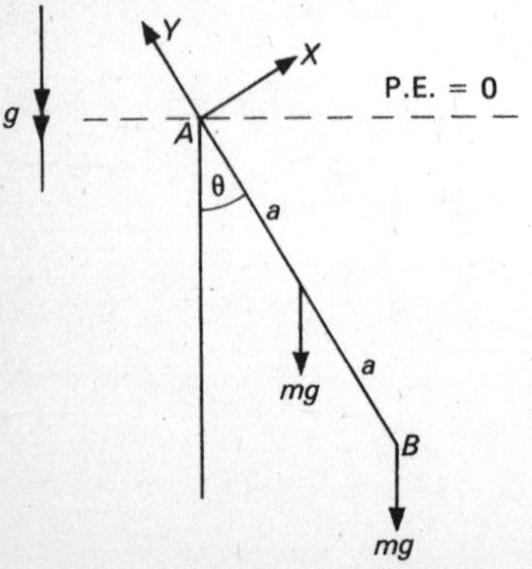

initially:

$\dot{\theta} = 0, \theta = \alpha$

moment of inertia of rod about $A = \frac{4}{3}ma^2$

moment of intertia of particle about $A = 4ma^2$

Figure 85

Refer to figure 85.

The reaction at the pivot does no work. The moment of inertia of the rod and the particle about the fixed end A

$$=\frac{4ma^2}{3}+4ma^2=\frac{16ma^2}{3}.$$

Applying **conservation of mechanical energy** to the system:

Initially,
$$\text{P.E.}=-mga\cos\alpha-2mga\cos\alpha$$
$$=3mga\cos\alpha$$
$$\text{K.E.}=0.$$

In the general position,

$$\text{P.E.}=-3mg\cos\theta$$
$$\text{K.E.}=\frac{1}{2}\frac{16}{3}ma^2\dot{\theta}^2.$$

Hence,
$$\dot{\theta}=\left(\frac{9g}{8a}(\cos\theta-\cos\alpha)\right)^{1/2}.$$

The acceleration of the centre of mass of a system, is that which would be produced by the resultant force acting on a particle of the total mass placed at the centre of mass of the system. The centre of mass in this case is distance $3a/2$ from A.

Resolving parallel to the rod:

$$Y-2mg\cos\theta=2m\dot{\theta}^2\frac{3}{2}a.$$

Substituting for $\dot{\theta}$,

$$Y=2mg\cos\theta+3ma\times\frac{9g}{8a}(\cos\theta-\cos\alpha)$$

when $\boldsymbol{\theta=0}°$,

$$Y=2mg+\frac{27}{8}mg-\frac{27}{8}mg\cos\alpha$$

$$=\mathbf{(43-27\cos\alpha)\frac{mg}{8}}.$$

Resolving perpendicular to the rod:

$$X - 2mg\sin\theta = 2m\frac{3}{2}a\ddot{\theta}.$$

Differentiating $\dot{\theta}$ w.r.t. time,

$$2\dot{\theta}\ddot{\theta} = -\frac{9g}{8a}\sin\theta.$$

Hence when $\boldsymbol{\theta = 0°}$, $\boldsymbol{X = 0}$.

In order to find the period of small oscillations, we apply '$C = I\ddot{\theta}$', i.e. 'rate of change of angular momentum about a fixed axis = resultant moment of the external forces about that axis'.

Taking moments about A in the direction of increasing θ,

$$-\frac{3a}{2}2mg\sin\theta = \frac{16}{3}ma^2\ddot{\theta}$$

$$\ddot{\theta} = -\frac{9g}{16a}\sin\theta.$$

We are considering small oscillations, thus, $\sin\theta \triangleq \theta$ and

$$\ddot{\boldsymbol{\theta}} = -\frac{\boldsymbol{9g\theta}}{\boldsymbol{16a}}$$

which is an **equation of motion for simple harmonic motion, of period**

$$\boldsymbol{T = \frac{8\pi}{3}\sqrt{\frac{a}{g}}}.$$

Example 2

Prove, by integration, that the moment of inertia of a uniform rod, of length $2a$ and mass M, about an axis perpendicular to the rod and passing through one end, is $4Ma^2/3$.

A body is made up of two equal uniform rods OA and CD, of mass M and length $2a$, which are joined together to form a letter **T**. The end A of the rod OA is rigidly attached to the midpoint of the rod CD, so that OA and CD are perpendicular.

The body is suspended from O, with A vertically below O, and is free to rotate in a vertical plane about a horizontal axis through O, perpendicular to the plane of the body. The point A is now given a velocity $\sqrt{(nga)}$, where n is a constant, in the direction CD. Show that, when OA makes an angle θ with the downward vertical, the angular speed of the body, $\dot{\theta}$, is given by

$$a\dot{\theta}^2 = \frac{18g}{17}\left(\cos\theta - 1 + \frac{17n}{72}\right).$$

Deduce the value of n for which the point A just reaches the position vertically above O. For this value of n, find the reaction at the axis when A is vertically above O.

(A.E.B.)

The first part of this question is standard book work (e.g. *Key Facts Passbook: Applied Mathematics*, p. 285) and will not be shown here.

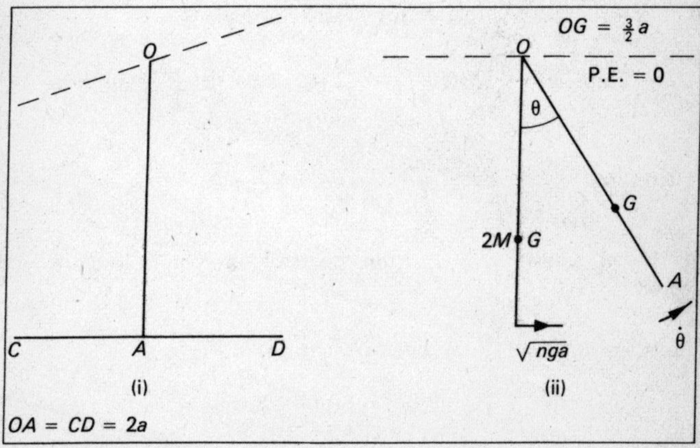

Figure 86

Refer to figure 86(i).

The moment of inertia of the rod OA about the said axis through O is

$$\frac{4Ma^3}{3}.$$

The moment of inertia of the rod CD about an axis through its mid point, perpendicular to the rod, is $\frac{1}{3}Ma^2$. By the **parallel axes theorem** the moment of inertia of the rod CD about the axis through O is

$$\tfrac{1}{3}Ma^2+(2a)^2M.$$

Thus, **the moment of inertia of the T-shape about the axis through O is**

$$\frac{4}{3}Ma^2+\frac{1}{3}Ma^2+4a^2M=\frac{17}{3}Ma^2.$$

Refer to figure 86(ii).

The centre of gravity, G, of the **T**-shape is distant $3a/2$ from O along OA. The mass of the body is $2M$.

The point A is given a velocity $\sqrt{nga}$ in the direction CD. **Mechanical Energy is conserved**. Thus, equating the M.E. initially with that in the general position shown, where OA makes an angle θ with the downward vertical,

$$-2Mg\frac{3}{2}a+\frac{1}{2}\frac{17}{3}Ma^2\frac{nga}{4a^2}=-2Mg\frac{3}{2}a\cos\theta+\frac{1}{2}\frac{17}{3}Ma^2\dot{\theta}^2.$$

Therefore, $$a\dot{\theta}^2=\frac{18g}{17}\left(\cos\theta-1+\frac{17}{72}n\right).$$

If A just reaches the position vertically above O, then when $\theta=180°$, $\dot{\theta}=0$.

Thus, substituting in the above equation,

$$0=\frac{18g}{17}\left(-1-1+\frac{17}{72}n\right)$$

$$n=\frac{144}{17}.$$

The **reaction at the axis** when A is vertically above O will be such that it will support the weight of the body, i.e. $\mathbf{2}\boldsymbol{Mg}$ **vertically upwards**.

Example 3

The moment of inertia of a lamina of mass M about a normal axis through its centre of gravity, G, is Mk^2 and a particle of mass M is attached to the lamina at A, where $AG=2h$. C is a point of the lamina such that AGC is an equilateral triangle. The lamina is free to rotate in a vertical plane about a smooth horizontal axis through C, perpendicular to the plane of the lamina. Show that small oscillations of the lamina about its position of stable equilibrium are simple harmonic and write down the length of the equivalent simple pendulum. (If a formula is used for results relating to a compound pendulum it must be proved.) Find the value of h such that the length of the equivalent simple pendulum should be least.

(S.U.J.B.)

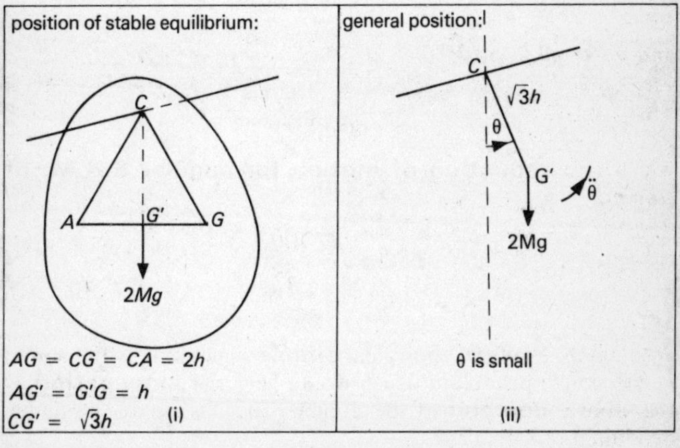

Figure 87

Refer to figure 87(i).

By the **Parallel Axis theorem**, the moment of inertia of the lamina (without the particle at A attached) about the axis through C, perpendicular to the plane of the lamina

$$=\boldsymbol{Mk^2+M4h^2}.$$

The moment of inertia, I, of the lamina with the particle of mass M attached at A is,

$$(Mk^2+M4h^2)+(M4h^2), \text{ i.e. } \boldsymbol{I=M(k^2+8h^2)}.$$

The position of stable equilibrium is where the centre of gravity, G', of the lamina with the attached particle, is vertically under the point of suspension, see figure 87 (i).

Refer to figure 87 (ii).

The lamina with the attached particle performs small oscillations about its stable equilibrium position. We shall examine the equation of motion for the general position shown.

Applying Newton's Law for rotation '$C=I\ddot{\theta}$',

$$-2Mg\sqrt{3}\,h\sin\theta=M(k^2+8h^2)\ddot{\theta}.$$

Thus,
$$\ddot{\theta}=\frac{-2\sqrt{3}\,hg}{(k^2+8h^2)}\sin\theta.$$

and for small θ, $\sin\theta \triangleq \theta$.

Therefore,
$$\ddot{\theta}=\frac{-2\sqrt{3}\,hg\,\theta}{(k^2+8h^2)}$$

which is an **equation of motion for angular S.H.M. of period**

$$\boldsymbol{T=2\pi\sqrt{\frac{k^2+8h^2}{2\sqrt{3}\,hg}}}.$$

The length, P, of the equivalent **simple** pendulum, is the length of the simple pendulum which would have the **same period** as the above **compound** pendulum. The period of a simple pendulum of length l is

$$2\pi\sqrt{\frac{l}{g}}.$$

Comparing these periods, we see that

$$\boldsymbol{l=\frac{k^2+8h^2}{2\sqrt{3}h}}.$$

Differentiating w.r.t. h,

$$\frac{dl}{dh}=\frac{2\sqrt{3}h\,16h-(k^2+8h^2)2\sqrt{3}}{12h^2}$$

when $\frac{dl}{dh}=0$, $\qquad h=\pm\frac{k}{2\sqrt{2}}$.

Since $\frac{d^2l}{dh^2}=\frac{2\sqrt{3}}{12}\left(\frac{2k^2}{h^3}\right)>0$ when $h=\frac{+k}{2\sqrt{2}}$,

then the **least value of l occurs when**

$$\boldsymbol{h=\frac{k}{2\sqrt{2}}}.$$

Example 4

A uniform circular disc, of mass m and radius a, is free to rotate about a horizontal axis through its centre, perpendicular to its plane. The disc is at rest when a particle of mass $\frac{1}{2}m$ is dropped on it from above. It strikes the circumference and sticks to it at a point Q, such that OQ makes an angle of 30° with the upward vertical through the centre of the disc. Find the angular speed of the compound body immediately after impact, in terms of the velocity, u, of the particle immediately before impact.

Show that the disc with the particle attached will perform complete revolutions if $u^2=8ga$.

The moment of inertia of the disc about the axis through O is $\frac{1}{2}ma^2$. The moment of inertia of the particle attached to the circumference of the circle is $ma^2/2$. Thus **the moment of inertia of the compound body about the horizontal axis through O is ma^2.**

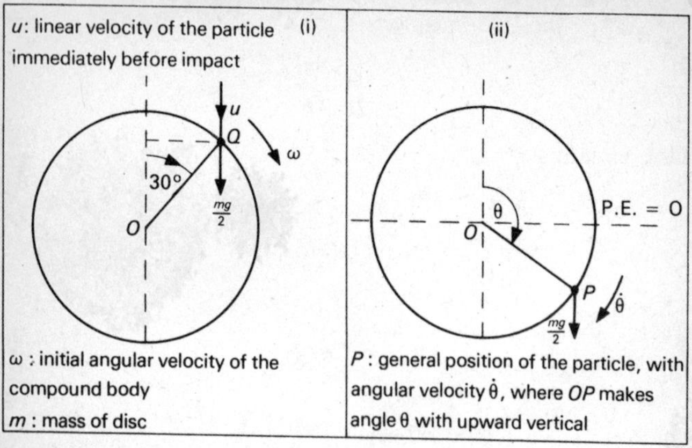

Figure 88

Refer to figure 88 (i).

At the impact of the particle on the circumference of the circular disc, the particle will impart an impulse to the disc, and the disc will impart an equal, opposite, impulse to the particle.

Hence the sum of the moments of the external forces about the axis is zero, and **angular momentum is conserved about the axis of rotation**.

Moment of linear momentum of the particle about the axis through O before impact

$$=\frac{1}{2}mua\cos 60°.$$

Moment of linear momentum of the particle after impact

$$=\frac{1}{2}m\omega a^2.$$

The angular momentum of the disc after impact is

$$\frac{1}{2}ma^2\omega.$$

Thus,

$$\frac{mua}{4}=\frac{ma^2\omega}{2}+\frac{ma^2\omega}{2}$$

and

$$\omega=\frac{u}{4a}.$$

(Alternatively, we could have considered the angular momentum of the **compound** body, $ma^2\omega$, after impact, and arrived at the same result).

Refer to figure 88 (ii).

After impact, no external forces are acting on the system, and therefore **mechanical energy is conserved.**

Initially,

$$\text{M.E.}=\frac{1}{2}mga\cos 30^\circ+\frac{1}{2}ma^2\frac{u^2}{16a^2}$$

$$=mga\frac{\sqrt{3}}{4}+\frac{mu^2}{32}.$$

In the general position, P,

$$\text{M.E.}=\frac{1}{2}mga\cos\theta+\frac{1}{2}ma^2\dot{\theta}^2.$$

By the Principle of Conservation of Mechanical Energy,

$$mga\frac{\sqrt{3}}{4}+\frac{mu^2}{32}=\frac{1}{2}mga\cos\theta+\frac{1}{2}ma^2\dot{\theta}^2$$

Thus,

$$\boldsymbol{a^2\dot{\theta}^2=\frac{u^2}{16}+ag\frac{\sqrt{3}}{2}-ag\cos\theta.}$$

The compound body will not perform complete revolutions if, for some value of θ, $\dot{\theta}=0$ at any stage.

When $\dot{\theta}=0$,

$$\frac{u^2}{16}+ag\frac{\sqrt{3}}{2}-ag\cos\theta=0.$$

When $u^2=8ag$, then for $\dot{\theta}=0$

$$\boldsymbol{\cos\theta=\frac{1+\sqrt{3}}{2}>1}, \text{ which is } \textbf{impossible}.$$

Therefore **the body performs complete revolutions for**

$$\boldsymbol{u^2=8ga}.$$

Example 5

A smooth narrow tube is in the form of a circle, of centre O and radius a, fixed with its plane vertical. Two particles A and B, both of mass m, are connected by a light inextensible string of length $\pi a/2$ and are placed in the tube with A at the highest point and the string taut. The system is released from rest. OA makes an angle θ with the upward vertical, at time t after release. Show that while the string remains taut,

$$a\dot{\theta}^2=g(1-\cos\theta+\sin\theta).$$

For this position of the particles, find the tension in the string and the force exerted by the tube on the particle A.

Refer to figure 89.

The tube is smooth and, after the system is released from rest, no external force does work (apart from gravity). Thus the system of forces is **conservative**, i.e. **Mechanical Energy is conserved**.

Initially,

$$\text{P.E.}=mga \qquad \text{K.E.}=0.$$

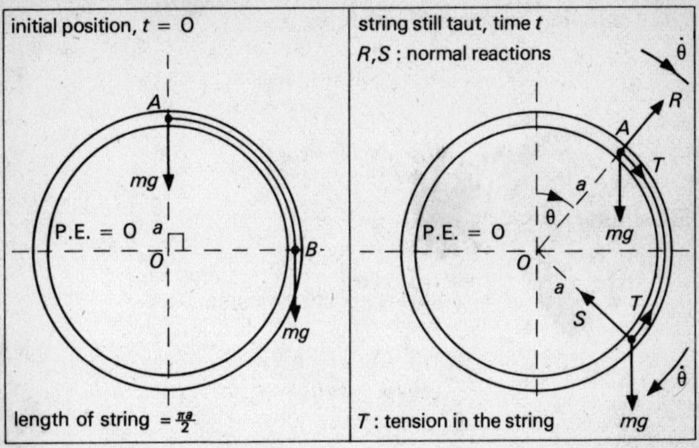

Figure 89

After time t,

$$\text{P.E.} = mga\cos\theta - mga\cos(90^\circ - \theta)$$
$$= mga(\cos\theta - \sin\theta)$$

and
$$\text{K.E.} = \frac{1}{2}mg\,\dot{\theta}^2 a^2 + \frac{1}{2}mg\,\dot{\theta}^2 a^2.$$

(While the string remains taut, both particles will move with the same angular speed, $\dot{\theta}$).

Equating,

$$a\dot{\theta}^2 = g(1 - \cos\theta + \sin\theta). \qquad \text{(i)}$$

Examining the equations of motion for A:

Radially,

$$R - mg\cos\theta = -m\dot{\theta}^2 a$$
$$\boldsymbol{R = mg(2\cos\theta - 1 - \sin\theta)}.$$

Tangentially, ('$f = ma$')

$$T + mg\sin\theta = m\ddot{\theta}a.$$

Differentiating equation (i), w.r.t. time,

$$2a\,\dot{\theta}\ddot{\theta}=g(\sin\theta\dot{\theta}+\cos\theta\dot{\theta})$$

$$\boldsymbol{a\ddot{\theta}=\frac{g}{2}(\sin\theta+\cos\theta).}$$

Substituting,

$$T=\frac{mg}{2}\sin\theta+\frac{mg}{2}\cos\theta-mg\sin\theta$$

$$\boldsymbol{=\frac{mg}{2}(\cos\theta-\sin\theta).}$$